Contents

1 Chapter 1: Early Electrical Pioneers

The story of electricity's journey from a scientific curiosity to a global necessity begins, not with sleek power stations, but with rudimentary devices and the ingenious minds who dared to dream of harnessing its power. Early experiments were far from efficient, often involving cumbersome static electricity generators or, later, primitive dynamos that produced only weak currents. These early pioneers, working with limited resources and understanding, faced countless setbacks. Yet, their relentless pursuit led to breakthroughs that would forever change the world. The initial generators, frequently based on Faraday's principles of electromagnetic induction, were unwieldy and inefficient, producing electricity in small amounts with considerable effort.

Imagine the challenges: crafting powerful magnets, developing durable conductors capable of handling the nascent electric current, and designing mechanical systems robust enough to drive the generators continuously. These weren't simply engineering problems; they were feats of innovation, demanding breakthroughs in materials science, mechanical engineering, and an understanding of electricity itself that was still in its infancy. Many of the early successes were driven by trial and error, careful observation, and an unwavering belief in the potential of this revolutionary force. Progress wasn't linear; it involved frequent setbacks, frustrating dead ends, and the slow accumulation of knowledge gained through painstaking experimentation.

The distribution of this newly generated electricity presented even more formidable obstacles. Early power grids, if they could even be called that, were often limited to a single building or a small neighborhood, constrained by the limitations of the technology. Think of thick, poorly insulated copper wires, prone to overheating and substantial power loss over even short distances. The concept of alternating current (AC) versus direct current (DC) was still a subject of intense debate, with Thomas Edison championing DC and George Westinghouse advocating for AC, a conflict that would shape the development of the electrical industry for years to come. This struggle reflected not only technological differences, but also fierce business competition and differing visions for the future of power distribution.

The early 20th century witnessed a significant shift. Advances in materials science produced stronger, more efficient conductors, and improvements in generator design yielded greater power output. The development of transformers allowed for the efficient transmission of electricity over longer distances, overcoming one of the major bottlenecks of early power grids. These improvements were not solely the result of individual genius; they were the product of collaborative effort, where scientists, engineers, and entrepreneurs built upon the foundational work of their predecessors. The rise of large-scale power plants marked a paradigm shift, indicating a move towards centralized generation and broader distribution networks.

But even with these advances, early power distribution remained unreliable and inefficient. Power outages were frequent, and the quality of the electricity delivered was often inconsistent. The cost of electricity remained high, limiting its accessibility to only the wealthy. The challenge was not just technical, but also economic and social. Getting electricity to the masses required not only technological innovation but also significant investment and widespread infrastructure development. This, in turn, prompted the development of better regulation, standards, and safety measures – crucial for the expansion and stabilization of this emerging sector. The growth wasn't simply

technological; it was fundamentally social and economic, shaping urban landscapes and industrial development in profound ways.
These early electrical pioneers, often working in relative obscurity, laid the groundwork for the sophisticated power systems we rely on today. Their contributions, frequently overlooked in the narrative of technological progress, were nothing short of transformative. They faced not only scientific and engineering challenges but also social and economic hurdles, overcoming skepticism, limitations of resources, and the constant pressure to prove the viability of their inventions. Their persistence and ingenuity ultimately created the foundation upon which the electrical industry would flourish, transforming societies and paving the way for the advancements explored in subsequent chapters. The legacy of these pioneering figures resonates even now, a testament to their vision and dedication. Their work stands as a powerful reminder that significant progress is often the result of incremental steps, perseverance, and a willingness to confront the unknown.

1.1 The First Generators

Early experiments with electricity were often rudimentary, involving static electricity and simple electrochemical cells. These provided small amounts of power, suitable only for limited demonstrations or small-scale experiments. The real breakthrough came with the development of the first practical generators, devices that could convert mechanical energy into electrical energy on a larger scale. These early machines weren't particularly efficient, and their output was relatively low, but they represented a giant leap forward in our ability to harness electricity.
One pivotal invention was the Faraday disc, developed by Michael Faraday in 1831. This device, a simple copper disc rotating between the poles of a magnet, demonstrated the principle of electromagnetic induction, a cornerstone of electrical generation. While not a practical generator in itself, the Faraday disc provided a crucial understanding of how electricity could be produced from mechanical

motion; a fundamental concept for future advancements. Its simplicity belied its profound impact, paving the way for more sophisticated designs.

Early generators were primarily direct current (DC) machines, meaning the electricity they produced flowed in one direction. This had significant limitations for long-distance transmission. DC current experienced high losses over distance, making it impractical for widespread distribution. Imagine the limitations of supplying electricity to even a modest-sized town. That inherent constraint in the initial generators spurred innovation and the pursuit of a more efficient and viable solution for transporting electricity.

The development of the dynamo, a more robust and efficient DC generator, marked a crucial turning point. Dynamos utilized multiple coils of wire rotating within a magnetic field, producing a larger and more consistent electrical current. They were more powerful than their predecessors and became essential in powering early industrial applications such as arc lighting and electric motors, gradually increasing the demand and the necessity of enhancing electricity generation. This period also saw the rise of notable inventors like Zénobe Gramme, whose Gramme dynamo, characterized by its ring armature design, was notable for its ability to provide a relatively smooth and stable current.

The late 19th century witnessed a race to improve generator technology. The quest for higher efficiency and greater power output drove inventors to experiment with different designs and materials. Competition among inventors fueled rapid advancements, leading to the development of more powerful machines capable of generating enough electricity to meet the growing demands of industry and urban areas. Every improvement spurred further refinement, escalating the pace of innovation.

The transition to alternating current (AC) generators proved to be transformative. Invented by inventors such as Galileo Ferraris and Nikola Tesla, AC generators offered significant advantages over their DC counterparts, especially for power transmission. AC electricity could be easily stepped up to higher voltages for efficient long-distance transmission, using transformers, and then stepped down to lower voltages for safer and more practical usage in homes and factories. This was a crucial step that allowed the electric grid to become widespread and reliable, transforming societies around the world.

The first large-scale power plants emerged, utilizing steam engines to drive generators. These early power plants, often located near coal mines or water sources for steam generation, represented a crucial integration of existing technologies. The combination of steam power and electricity generation marked a pivotal moment in history, enabling the widespread use of electricity and greatly influencing societal development. The early power plants, though initially small by today's standards, were revolutionary in their ability to provide a constant, reliable supply of electricity to cities and industries. The early days of large-scale power generation laid the groundwork for future developments and the continued expansion of electricity grids worldwide. This era not only produced the first large-scale power plants, it also established many of the safety and operational standards that remain relevant today.

The story of the first generators is one of ingenuity, perseverance, and relentless innovation. From humble beginnings with static electricity and simple electrochemical cells to the powerful AC generators that power our world today, the journey showcases the incredible progress made in harnessing and utilizing electricity. The initial hurdles encountered in efficiency, transmission, and scale were overcome through sustained effort, resulting in a technology that fundamentally transformed human civilization. The challenges and successes of these early pioneers serve as a testament to the power of human creativity and the enduring impact of technological advancement. The development of these early generators and power systems not only supplied electricity, but it helped spur further developments in other technologies and fields, transforming the world in unexpected ways.

1.2 Early Power Distribution

Early electrical distribution systems faced immense challenges. Imagine the limitations: power generation was localized, often confined to a single building or a very small area. Transmission was rudimentary at best, relying on thick, inefficient copper wires strung along poles, susceptible to damage and significant energy loss through resistance. The distances over which electricity could be practically transmitted were severely restricted, confining the benefits of this new technology to a limited radius. This constrained growth and hindered widespread adoption.

The initial designs were incredibly simple, far removed from the sophisticated networks we have today. Direct current (DC), the primary form of electricity generated in these early days, proved highly inefficient for long-distance transmission. Voltage drops over distance meant that power rapidly dissipated, rendering even modest distances impractical without massive, and expensive, copper conductors. This severely limited the potential reach of early power stations.

Furthermore, the lack of efficient switching and control mechanisms made it difficult to manage the flow of electricity. Fault detection and protection were rudimentary, increasing the risk of outages and potential hazards. System reliability was consistently a major issue, causing frequent disruptions and adding to public concerns and hesitation about embracing this nascent technology.

Thomas Edison's early DC systems, while revolutionary, illustrated these constraints clearly. His Pearl Street Station in New York City, while a landmark achievement, served a very limited area. The high cost of copper and the limitations of DC transmission meant that expanding the network was extraordinarily expensive and complex, significantly limiting its growth potential. This highlights a crucial aspect of early power distribution—the fundamental conflict between scalability and practicality.

The invention and subsequent refinement of the transformer represented a watershed moment. This device, capable of efficiently stepping up and stepping down voltage, proved to be the key to unlocking the potential of alternating current (AC) for long-distance transmission. Lower currents at higher voltages meant dramatically reduced power loss over distance, allowing power to travel much further with far less energy waste. This fundamental shift paved the way for more extensive power grids.

The adoption of AC also facilitated the use of more efficient motors and other electrical equipment, further boosting the practical applications of electricity. The increased reliability and range of AC power helped overcome the inherent limitations of early DC systems, driving the rapid expansion of electrical networks across cities and eventually, countries. This evolution demonstrates the power of technological breakthroughs in transforming an entire industry.

However, even with the advent of AC, the early distribution systems faced considerable challenges. The standardization of voltage and frequency was a lengthy and complex process, necessitating significant coordination between different power companies and regions. Without such standards, interconnectivity was impossible, hindering the development of truly expansive power grids.

Moreover, the construction and maintenance of extensive power lines across vast geographical areas posed significant logistical and engineering challenges. This required major investment in infrastructure, skilled labor, and robust safety protocols. The widespread adoption of electricity was only possible with significant government support and private investment in the developing power infrastructure.

The early days of electricity distribution, therefore, weren't just about the technology itself. It was about the intricate interplay of technological innovation, economic viability, political will, and regulatory frameworks. These factors, working in tandem, were essential in propelling electricity from a localized novelty to a ubiquitous necessity that shapes our world today. The story of early power distribution is a testament to human ingenuity, resilience, and the collective effort required to transform a revolutionary concept into a reality.

2 Chapter 2: The Rise of Steam Power

The Industrial Revolution's insatiable hunger for power found a potent ally in steam. Before electricity's widespread adoption, steam engines, initially used in mines and factories, began powering generators, marking a pivotal shift in energy production. This wasn't simply a matter of replacing one power source with another; it was a fundamental leap forward in scale and efficiency. Early steam engines, clunky and inefficient, laid the groundwork for the sleek, powerful turbines that would dominate power generation for a century.
The early power plants that emerged were marvels of engineering for their time. Imagine massive structures, filled with the rhythmic chugging of steam engines, the hiss of escaping pressure, and the relentless turning of generators. These weren't the pristine, automated facilities of today; they were raw displays of industrial might, demanding constant monitoring and manual labor. Workers, often in harsh conditions, were essential to their operation, highlighting the human cost alongside the technological advancement. These plants, however rudimentary compared to modern standards, represented a significant step towards centralized power generation, delivering electricity to growing urban centers.
Improving steam engine efficiency became an obsession for engineers. Minute adjustments in valve timing, the optimization of boiler pressure, and the refinement of turbine blade designs all yielded significant gains. The quest for greater thermal efficiency drove innovation,

leading to increasingly sophisticated designs. The development of the compound steam engine, which used multiple expansion stages to extract more energy from the steam, stands as a testament to this relentless pursuit of optimization. This wasn't simply about increasing output; it was about reducing fuel consumption, making power generation more economically viable and environmentally responsible, even in the context of the time. The race to enhance efficiency was fueled by both economic incentives and the inherent drive to push the boundaries of engineering capabilities.

The advent of the steam turbine represented a profound technological breakthrough. Unlike reciprocating steam engines, which employed pistons moving back and forth, turbines used the kinetic energy of high-velocity steam to rotate a shaft directly, leading to smoother operation and higher rotational speeds. This allowed for larger generators, capable of producing far greater amounts of electricity. The turbine's compact design compared to earlier steam engines also facilitated construction of more efficient power plants. This compact design also allowed for higher rotational speeds, leading to increased power generation from smaller spaces. The development of the steam turbine irrevocably altered the landscape of power generation, paving the way for the enormous power plants of the 20th century.

However, this rise wasn't without its challenges. The sheer scale of early steam power plants necessitated significant infrastructure development. Coal, the primary fuel source, needed to be transported in vast quantities, requiring extensive rail networks and port facilities. The plants themselves demanded skilled labor for operation and maintenance. And while steam power revolutionized energy production, the inherent inefficiencies and environmental impact of burning fossil fuels – even in this early phase – began to cast a long shadow, a problem that would only grow more significant in the following decades. This early period shows that while innovation propels progress, it also gives rise to new sets of problems that need to be solved. The seeds of future advancements were already sown, even at the height of the era of steam power.

The story of steam power's rise is not simply a tale of technological advancement; it's a narrative of human ingenuity, economic drivers, and the gradual awareness of environmental consequences. It underscores the iterative nature of progress, where each step, however significant, lays the groundwork for further refinement and evolution. The advancements made in steam power laid a crucial foundation for subsequent breakthroughs in electrical generation, providing a framework and a precedent for efficiency, scale, and the relentless pursuit of better ways to harness energy. The lessons learned from this era continue to resonate in modern power generation, highlighting the enduring importance of efficiency, resource management, and the inevitable trade-offs inherent in harnessing energy to power our civilization.

2.1 Steam Turbine Technology

Steam turbines represent a cornerstone of large-scale electricity generation, their development intrinsically linked to the expansion of power grids and industrialization. Early steam turbines, while significantly improving upon reciprocating engines, were relatively inefficient and prone to mechanical failures. The inherent limitations of early designs, such as blade erosion and material fatigue, hampered their widespread adoption. However, ongoing research and development addressed these issues, gradually increasing efficiency and reliability. This iterative process, fueled by the demands of burgeoning industries, propelled the evolution of steam turbine technology. Their adoption marked a clear step towards generating power at a scale previously unimaginable.

The advent of improved materials science significantly impacted steam turbine design. The introduction of stronger, heat-resistant alloys enabled the construction of turbines capable of operating at higher temperatures and pressures. These advancements directly translated to increased power output and improved thermal efficiency. This continuous improvement cycle demonstrates the crucial interplay between material science and the evolution of power generation technologies. Moreover, the understanding of thermodynamics and fluid dynamics advanced significantly in parallel with steam turbine design. These advancements in fundamental scientific principles allowed engineers to refine designs, optimizing blade geometry and steam flow for maximum efficiency. This interdisciplinary approach has been critical in advancing steam turbine technology to the impressive efficiency levels observed in modern designs.

Consider the impact of the Parsons reaction turbine, a revolutionary design that significantly improved efficiency and power output. This innovation, among others, laid the groundwork for the immense steam power plants that power much of our modern world. The design itself was revolutionary for its time, its precise engineering a marvel of its era. Subsequent iterations and refinements on this original design continue to shape the landscape of modern power generation. The development story isn't just about technological innovation. It's also a narrative of economic incentives and engineering ingenuity working in tandem. The demand for more efficient and reliable power generation fueled intense competition and innovation, leading to rapid advancements in steam turbine technology. This continuous cycle of innovation demonstrates the power of market forces in shaping technological progress.

Moreover, the ongoing development of steam turbines isn't confined to simple efficiency improvements. Engineers continue to grapple with challenges such as reducing emissions and improving the overall sustainability of steam power plants. The integration of advanced control systems and the use of renewable energy sources as a heat source for the steam cycle reflect the evolving priorities in the energy sector. Furthermore, the exploration of novel turbine designs, using different materials and operating principles, aims to increase efficiency and lower the environmental impact, indicating a continuing push towards sustainable energy solutions. This ongoing pursuit highlights the ever-evolving nature of power generation technologies and adaptation to broader societal demands.

In conclusion, steam turbine technology has undergone a remarkable journey, transforming from relatively inefficient early designs to highly sophisticated and efficient systems that form the backbone of many large-scale power plants today. The continuous innovation in materials science, thermodynamics, and engineering practices has propelled this evolution, demonstrating the crucial interplay between scientific advancement and technological progress. It's a continuous story, far from its conclusion, with ongoing research pushing the boundaries of efficiency, sustainability, and ultimately, our capacity to harness energy and power the world. Understanding this journey provides insight into the dynamism of power generation and the ever-present drive for improved performance and environmental responsibility. The development of steam turbine technology serves as an excellent case study of how engineering ingenuity, driven by societal needs, can lead to remarkable technological leaps and profound transformations in how we harness and utilize energy.

2.2 Early Power Plants

Early steam-powered electricity generation wasn't a straightforward process; it was a messy, iterative affair, filled with trial and error, explosions, and breakthroughs. The first power plants weren't sleek, modern facilities; they were often cobbled-together affairs, reflecting the nascent understanding of both steam technology and electrical principles. Think of noisy, inefficient behemoths, belching smoke and steam, far from the pristine image we often associate with power generation today. These early plants relied heavily on reciprocating steam engines, a technology already mature in other industrial sectors, but their adaptation to electricity generation presented unique challenges. The inherent pulsating nature of these engines required significant engineering ingenuity to convert the intermittent power into a relatively smooth electrical output.

Early designs often involved a bewildering array of belts, pulleys, and gears, creating a cacophony of clanking metal and a constant risk of mechanical failure. Maintenance was a constant headache, and downtime was frequent. The conversion of the steam engine's rotary motion into the electrical current was also a major obstacle. Early dynamos, the direct current (DC) generators of the time, were themselves unreliable and prone to overheating. The lack of efficient, high-voltage transmission lines meant that these power plants could only supply electricity to a very limited radius, further hindering their impact.

Despite these limitations, these early power plants represented a giant leap forward. They proved the feasibility of generating electricity on a scale larger than ever before seen, paving the way for the electrification of cities and industries. The experience gained from these early, often frustrating attempts, was crucial in shaping the design and engineering of the far more efficient and reliable power plants that followed. The fundamental principles of steam power generation—converting heat energy from burning fuel into mechanical energy, which then drives a generator to create electricity—were established during this era. These early plants, though rudimentary by modern standards, laid the foundation for a global energy infrastructure.

One of the most significant challenges faced in these early plants was the efficiency of energy conversion. Early steam engines were far from perfect; much of the heat energy from burning coal was lost as waste heat, leading to incredibly low overall efficiency. This meant that vast amounts of fuel were required to generate even a modest amount of electricity, driving up costs and significantly impacting environmental pollution. This low efficiency also dictated the close proximity of power plants to their consumers, as the energy losses during transmission over longer distances would have been prohibitive. This geographical limitation became a key factor in urban planning during the early stages of industrial electrification. The concentration of power plants near population centers often led to significant air pollution problems in urban areas.

The construction of these early power plants also involved significant engineering and logistical challenges. The scale of the machinery required was unprecedented, necessitating novel approaches to building design and material handling. Often, these power plants were situated near sources of fuel, such as coal mines or rivers for water power, which dictated the location of the infrastructure. The workforce also

needed to adapt to the demands of a new technology, developing expertise in the operation and maintenance of steam engines, dynamos, and the emerging electrical grid. Learning curves were steep, and accidents were frequent, adding a layer of difficulty and risk to the enterprise.
The early power plants were not just technical marvels; they were also symbols of societal progress. The ability to generate electricity on a large scale was transforming industries and cities, fundamentally changing the way people lived and worked. However, this progress came at a cost. The environmental impact of burning vast quantities of coal was already becoming apparent in the early decades of the 20th century. The consequences of atmospheric pollution and its effect on human health were beginning to be understood, paving the way for the future push towards cleaner and more sustainable forms of energy generation. The early power plants were a crucial step, though a somewhat clumsy one, towards a technologically advanced future.

2.3 Efficiency Improvements

Early steam power plants were notoriously inefficient. Much of the energy generated from burning coal was lost as heat escaping into the atmosphere. Improving this was paramount to making steam power a truly viable and scalable energy source. Engineers focused on optimizing the entire process, from boiler design to turbine efficiency. The development of more efficient boilers meant less fuel was needed to produce the same amount of steam, directly impacting operating costs and reducing environmental impact.
Improvements to turbine design were equally crucial. Early turbines suffered from significant frictional losses, reducing their overall output. Innovations like improved blade materials and more aerodynamic designs minimized these losses, leading to a notable increase in power generation for the same amount of steam input. Furthermore, advances in condenser technology allowed for better recovery of waste heat, improving the overall thermodynamic cycle efficiency.

These seemingly small changes drastically altered the landscape of electricity generation.

Careful attention to the entire thermodynamic cycle, rather than just individual components, yielded significant advancements. This holistic approach involved analyzing and improving heat transfer processes at every stage, including preheating feedwater and optimizing the pressure and temperature conditions throughout the cycle. A better understanding of thermodynamics principles allowed engineers to refine the design of steam cycles, incorporating advanced techniques like reheating and regeneration to significantly increase the overall efficiency of the plants. This optimization didn't just affect the bottom line; it also decreased the environmental footprint of these facilities by significantly reducing fuel consumption.

The development of high-pressure, high-temperature steam technology represented a considerable leap forward. Pushing the boundaries of material science, engineers created turbines and boilers capable of operating under significantly more demanding conditions, far exceeding the capabilities of earlier designs. This increased operational pressure and temperature yielded much higher energy density in the steam, leading to a substantial increase in power output. This development was not without challenges. Dealing with the high pressures and temperatures required the design and implementation of advanced safety systems and more robust materials, a significant technological hurdle in itself.

Alongside technological improvements, operational practices were refined. Regular maintenance schedules became critical in sustaining peak efficiency. This required better monitoring systems and the development of predictive maintenance strategies, identifying potential issues before they led to significant performance degradation or costly downtime. Improved operator training also played a vital role, ensuring that plants operated consistently and efficiently. These subtle changes, although not directly related to technological development, were still crucial in optimizing the overall efficiency of power generation.

Another significant advancement was the introduction of better control systems for steam plants. Early systems were often basic and relied heavily on manual adjustments. The introduction of sophisticated automatic control systems enabled real-time monitoring of various plant parameters. Automated control systems, using feedback loops and advanced algorithms, allowed for precise adjustments to parameters such as steam pressure, temperature, and flow rates, ensuring that the plant operated at its optimum efficiency at all times. These systems significantly reduced the likelihood of human error and optimized plant performance across various operating conditions.

The pursuit of efficiency also led to the development of combined cycle power plants. This approach combined the efficiency of gas turbines with steam turbines in a single plant. The hot exhaust gases from the gas turbine are used to generate steam, which then drives a steam turbine, recovering additional energy that would have otherwise been lost. This synergistic approach significantly increased overall efficiency compared to using either turbine type alone, setting new standards in the industry. The implementation of combined cycles highlighted the importance of integrating different technologies to enhance overall performance, showcasing a move towards systems thinking in power generation. This innovation was a testament to the evolution of not just individual components, but also of the overall system design.

Beyond the immediate technological improvements, a fundamental shift occurred within the industry. The focus shifted from simply increasing power output to maximizing efficiency. This change reflected a deeper understanding of the economics and environmental impact of power generation. A more efficient plant meant lower fuel costs, reduced emissions, and a smaller environmental footprint. This paradigm shift drove further innovation and shaped the future direction of power plant design and operation. This marked a transition from a purely production-focused approach to a more sustainable and cost-effective one.

The impact of these efficiency improvements is undeniable. Steam power plants have transitioned from relatively inefficient early designs to highly optimized systems capable of generating electricity with significantly higher efficiency. This improvement didn't happen overnight. It was a gradual process driven by continuous technological innovation and a deeper understanding of thermodynamics. The advancements in materials science, control systems, and plant design all played critical roles in the overall increase in efficiency, allowing steam power to remain a major player in the global energy landscape. The legacy of these improvements continues to inform and influence modern power generation technologies.

3 Chapter 3: Harnessing Water Power

Harnessing the immense power of moving water has been a cornerstone of electrical generation for over a century. Hydroelectric dams, those monumental structures spanning rivers and gorges, represent a powerful and often overlooked source of clean energy. Their construction, however, is a complex undertaking, demanding meticulous planning and significant engineering expertise. From initial site surveys to the final stages of commissioning, every step involves careful consideration of geological factors, environmental impact, and overall feasibility. Choosing the right location is paramount; a sufficient head of water, a reliable water source, and appropriate geological conditions are all crucial.

The design process itself is an intricate blend of art and science. Engineers must carefully assess the water flow, considering variations in seasonal rainfall and potential flood events. The dam's structure itself is a marvel of engineering, typically comprising concrete, rock, or earth, designed to withstand immense water pressure and the forces of nature. Turbines, the heart of the hydroelectric power generation system, are strategically placed within the dam's structure to harness the kinetic energy of the falling water. These powerful machines convert the water's movement into mechanical energy, which is then used to drive generators, producing electricity.

Hydropower efficiency, while generally high, is not without limitations. Factors such as turbine design, water head, and overall system losses all play a role in determining the overall efficiency. Modern advancements in turbine technology, particularly the development of more efficient designs, have significantly improved energy conversion rates. Furthermore, optimization of water flow within the system can minimize losses and maximize output. Innovative techniques, such as advanced control systems and real-time monitoring, are improving overall efficiency while reducing environmental impact.

Yet, the environmental implications of hydroelectric dams cannot be ignored. The construction of a dam inevitably alters the natural river ecosystem, affecting downstream water flow, aquatic habitats, and sediment transport. Large-scale dams can lead to habitat loss for fish and other aquatic species, disrupting ecological balance. Mitigation strategies, including fish passages and other environmental protection measures, are essential to minimize these impacts. Careful consideration of these factors during the planning and design phases is critical for environmentally responsible hydropower development. Ongoing research and development are focused on sustainable practices to reduce the environmental footprint of hydropower generation.

Modern hydroelectric projects are increasingly integrating sustainable design principles. The focus is not only on maximizing power output but also on minimizing environmental damage. Projects today often incorporate features like fish ladders, designed to allow migrating fish to bypass the dam, helping maintain fish populations. Furthermore, the integration of renewable energy storage solutions with hydropower systems is gaining popularity, ensuring a more reliable and sustainable energy supply. This advancement is increasingly important to accommodate fluctuating demand and minimize reliance on fossil fuels.

The future of hydropower holds immense promise, with ongoing research focused on increasing efficiency, reducing environmental impact, and developing new hydropower technologies. The development of smaller-scale, run-of-river hydropower plants, for instance, offers an alternative to large-scale dams, minimizing environmental disruption while still providing a valuable source of renewable energy.

Furthermore, innovative designs and technologies are being explored to enhance the sustainability and efficiency of existing hydropower systems. As the world transitions towards a cleaner energy future, hydropower will continue to play a vital role, offering a clean, reliable, and sustainable energy source for generations to come. The key lies in striking a balance between the need for energy and the preservation of our natural environment. This requires not only technological innovation but also a commitment to sustainable practices, ensuring that the benefits of hydropower are enjoyed without compromising environmental integrity. The future demands a conscientious approach to hydroelectric development, combining technological innovation with responsible environmental stewardship.

3.1 Hydroelectric Dams

Hydroelectric dams represent a significant milestone in harnessing nature's power for electricity generation. Their construction, however, is a complex undertaking requiring meticulous planning, advanced engineering, and considerable financial investment. The process begins with a comprehensive site assessment, carefully evaluating factors like water flow, geological stability, and environmental impact. Detailed hydrological studies are essential to predict water availability throughout the year, ensuring consistent power output. This careful evaluation minimizes risk and maximizes the dam's long-term viability.

The actual dam construction itself is a monumental feat of engineering, involving the careful placement of massive concrete structures, often requiring years of labor. Sophisticated excavation techniques are employed to prepare the foundation, ensuring stability and longevity. The design of the dam integrates spillways, ensuring excess water can safely bypass the dam during periods of high rainfall, thus preventing catastrophic failure. Precise calculations and rigorous testing of materials are paramount in the construction process. These are critical considerations for the safety and successful operation of the dam.

Turbines and generators are the heart of a hydroelectric power plant, converting the potential energy of the water into usable electricity. The water flows through carefully designed penstocks and then impacts the turbine blades, causing them to spin at high speed. This rotational energy drives the generators, producing electricity that feeds into the regional power grid. The efficiency of this process hinges on optimal turbine design and efficient generator technology; modern hydroelectric plants are remarkably efficient, achieving high conversion rates of potential energy to electrical energy. Ongoing research focuses on improving these conversion rates.

Environmental concerns are inextricably linked to hydroelectric dam projects. The construction of large dams often leads to habitat destruction, altering river ecosystems significantly. This affects downstream habitats and could lead to various issues, potentially causing the relocation or extinction of certain species. The alteration of water flow patterns can negatively affect aquatic life and the surrounding ecosystem. Mitigation strategies, such as fish ladders and improved water management practices, are implemented to minimize the negative impacts, but these only go so far in alleviating the broader consequences.

In recent years, innovative designs have attempted to minimize environmental disruptions. Run-of-river hydropower plants, for example, avoid the creation of large reservoirs, reducing environmental impact, however, they do not supply a constant stream of energy. These smaller scale plants generate power by diverting a portion of the river's flow through turbines. This innovative design significantly reduces the environmental disruption associated with traditional hydroelectric dams. While less impactful, run-of-river designs also have a smaller power output. Balancing environmental responsibility with power production remains a critical challenge for hydroelectricity.

The future of hydropower is intertwined with sustainable development goals. Increasing global energy demands necessitate expanding renewable energy sources, and hydropower remains a pivotal player, albeit with challenges to overcome. Technological advancements in turbine technology, improved dam designs, and refined environmen-

tal mitigation strategies are vital for a sustainable future for hydroelectric power. Innovations focusing on increasing efficiency and reducing environmental impacts are key to ensuring the role of hydropower in the future energy mix. A future focused on sustainable power generation must consider the comprehensive approach of minimizing the downsides while maximizing benefits.
Economic considerations play a critical role in the feasibility of hydropower projects. While initial investment costs are substantial, long-term operational costs are relatively low, making hydroelectric power a cost-effective energy source in many locations. The long lifespan of dams, often exceeding several decades, contributes to the economic viability of these projects. However, the upfront investment and maintenance are not insignificant factors and are often dependent on the scale of the project. A careful cost-benefit analysis is necessary before undertaking the construction of any hydroelectric dam.
The social implications of large-scale hydroelectric dam projects are far-reaching, potentially displacing communities and affecting cultural heritage. Careful consideration must be given to the social aspects of dam projects; this can lead to many complications, therefore community engagement and the development of resettlement plans are vital to ensure the project's social acceptability and to minimize displacement. Transparency and collaboration with impacted communities are essential for responsible hydropower development. Many social considerations must be accounted for before undertaking any large-scale dam projects.

3.2 Hydropower Efficiency

The efficiency of a hydroelectric power plant hinges on several key factors, starting with the design of the dam itself. A well-designed dam maximizes the head, or the vertical distance between the water's surface and the turbine, directly impacting the potential energy available for conversion to electricity. The geometry of the dam's spillways and intake structures also plays a crucial role, influencing water flow and minimizing energy losses due to turbulence or friction. Careful consideration of these elements can significantly improve the overall efficiency of the hydropower system.

Turbine selection is paramount. Different turbine types, such as Francis, Kaplan, and Pelton turbines, are suited to varying water flow rates and heads. Choosing the wrong turbine for a particular site can lead to substantial energy losses. Modern turbines, incorporating advanced designs and materials, boast significantly higher efficiencies than their predecessors. Furthermore, regular maintenance and timely upgrades are essential to maintain optimal turbine performance and prevent efficiency degradation over time. Neglecting maintenance can lead to significant energy losses and reduced lifespan of the equipment.

The generator itself is another critical component. High-efficiency generators minimize losses during the conversion of mechanical energy from the turbine into electrical energy. Factors like the generator's design, materials, and cooling system all contribute to its efficiency. Modern generators often utilize advanced technologies, such as superconducting materials, to further minimize losses and boost overall output. Regular inspections and maintenance of the generator are crucial for maintaining its peak efficiency and ensuring a long service life. This involves monitoring for signs of wear, tear and potential issues that could impact performance.

The entire system's efficiency isn't just about individual components; it's about the synergy between them. This includes the penstocks (pipes carrying water to the turbines), which should be designed to minimize friction losses. The transmission lines, responsible for carrying the generated electricity to the grid, also play a role; energy losses during transmission can be significant, and using advanced transmission technologies can greatly mitigate these losses. Investing in these seemingly minor aspects of the system often brings disproportionately large returns in terms of overall efficiency improvement.

Beyond the physical infrastructure, operational strategies significantly influence hydropower efficiency. Careful water management, including optimizing water flow rates to match the demand, prevents energy waste. Moreover, efficient monitoring and control systems allow for real-time adjustments, maximizing energy generation based on fluctuating demands and available water resources. Sophisticated control systems and predictive modeling can help anticipate peak demands and optimize water release from the reservoir to maximize power generation during those periods.

Environmental factors also influence hydropower efficiency. Sedimentation in the reservoir can reduce the effective head and impair turbine performance. Regular dredging operations help maintain optimal water levels and ensure continued high efficiency. Similarly, changes in water temperature or flow rates due to climate change can affect the overall efficiency of the hydropower plant. Adapting to these factors and implementing mitigation strategies is crucial for long-term performance and sustainable energy production. This includes exploring alternative approaches to maintaining water flow in times of drought or other climatic abnormalities.

Finally, advancements in technology are constantly improving hydropower efficiency. Research into new turbine designs, advanced control systems, and energy storage solutions is continuously pushing the boundaries of what's possible. Staying abreast of these advancements and considering their implementation is essential for operators looking to maximize energy production and minimize environmental impact. Investing in research and development, not just in the hardware, but also in optimization strategies, will bring long-term returns to any hydropower plant. Furthermore, collaborative efforts between engineers, scientists, and policymakers can unlock further innovations and accelerate the transition towards a more sustainable energy future. This collaborative effort is key to ensuring efficient and environmentally responsible use of this valuable renewable resource.

3.3 Environmental Impacts

Hydroelectric power, while a clean energy source compared to fossil fuels, carries significant environmental consequences. The construction of large dams fundamentally alters river ecosystems. The natural flow of rivers is disrupted, impacting fish migration patterns, sediment transport, and the overall biodiversity of the riverine environment. This disruption can lead to habitat loss for numerous aquatic species, potentially driving some towards extinction. Furthermore, the impounding of water behind dams creates large reservoirs that can lead to significant habitat loss in the flooded areas. This loss is not limited to aquatic life; terrestrial ecosystems are also impacted as forests and other habitats are submerged.

The alteration of water flow downstream of a dam can also have profound effects. Reduced water flow can affect downstream ecosystems, impacting riparian vegetation and altering the natural processes of erosion and sedimentation. The cooler water released from the depths of a reservoir can also negatively influence the temperature-sensitive aquatic life downstream. These changes can trigger cascading effects, disrupting the entire food web and affecting the health of the overall ecosystem. It's not simply a matter of immediate displacement; the long-term consequences are often more devastating, impacting ecological balance for decades or even centuries.

Consider the impact on water quality. Reservoirs, while appearing clean, can actually experience a decline in water quality due to decomposition of submerged organic matter and the accumulation of pollutants. These changes can lead to eutrophication, a process where excessive nutrients cause algal blooms, depleting oxygen levels and harming aquatic life. The release of water from reservoirs can also carry these pollutants downstream, contaminating river water and further stressing downstream ecosystems. Therefore, careful water quality monitoring and mitigation strategies are crucial during the planning and operation phases of hydroelectric projects.

Beyond the immediate environmental effects, there are indirect impacts to consider. The construction of dams can lead to habitat fragmentation, isolating populations of plants and animals and limiting their ability to reproduce and adapt. This isolation can make them more vulnerable to disease and environmental changes, further stressing biodiversity. The displacement of human populations due to dam construction is another significant consideration, especially for indigenous communities who have long-standing ties to the rivers and surrounding land. The social and cultural impacts of such displacements are often severe and long-lasting.

However, it is important to acknowledge that hydropower, in comparison to many other energy sources, offers several advantages. It is a renewable energy source, meaning it's not depleted through use and helps avoid reliance on fossil fuels. Hydropower plants also emit few greenhouse gases directly during operation, contributing less to climate change compared to fossil fuel-based power plants. Therefore, the aim shouldn't be to completely eliminate hydroelectric power but rather to develop more sustainable approaches to minimize environmental damage. This requires careful planning, thorough environmental impact assessments, and the implementation of mitigation strategies to reduce the impact on ecosystems and surrounding communities. Technological advancements in dam design, including fish ladders and improved water management strategies, are also crucial to lessen the negative environmental effects.

The key to responsible hydropower development lies in a comprehensive and holistic approach. This includes rigorous environmental impact assessments that fully evaluate the potential consequences of dam construction on both the aquatic and terrestrial environments. Moreover, effective mitigation measures must be implemented to minimize negative impacts. This could involve the creation of fish passageways, the restoration of riparian habitats, and the implementation of strategies to maintain water quality. The effective involvement of local communities in the planning process is crucial, not only to address their concerns but to foster responsible and sustainable projects that benefit both people and the environment. Ultimately, the aim is to balance the need for energy with the preservation of valuable ecological resources, fostering a future where clean energy production does not come at the cost of ecological damage. Ongoing research and development of innovative approaches are essential to improve the sustainability of hydropower projects. This includes exploring alternative dam designs, investigating ways to improve water management and fish migration, and continuously evaluating the environmental impacts of ongoing projects. Careful consideration of these factors is crucial to ensure the long-term health of ecosystems and to achieve a responsible balance between energy needs and environmental protection.

3.4 Modern Hydro Projects

Pumped hydro storage, a fascinating advancement, utilizes excess energy from other sources—like solar or wind—to pump water uphill to a reservoir. When demand increases, this stored water is released, driving turbines and generating electricity on demand. This system proves invaluable for balancing intermittent renewable energy sources, smoothing out fluctuations in supply and demand, and bolstering grid reliability. Its scalability, adaptable to varied terrains, and long operational lifespan make it a significant contributor to modern energy grids worldwide. The technology is constantly evolving, focusing on improved efficiency and reduced environmental impact.

Run-of-river hydroelectricity represents a less impactful approach compared to large dam projects. It harnesses the natural flow of rivers without significantly altering the river's course or creating extensive reservoirs. This method minimizes the disruption of aquatic ecosystems and reduces greenhouse gas emissions associated with large-scale reservoir creation. However, its energy output is directly linked to the river's flow rate, which can be inconsistent. Therefore, effective management and integration with other renewable sources are crucial for a reliable energy supply. Advancements focus on improving turbine efficiency and optimizing energy capture within the limitations of the natural river flow.

Marine current turbines, a relatively new technology, tap into the kinetic energy of ocean currents. These underwater turbines are designed to withstand the harsh marine environment and generate electricity from the continuous movement of seawater. Though still in its developmental stages, this technology offers enormous potential, particularly in areas with strong currents. Research and development focus on improving turbine design for efficiency and durability, addressing challenges like corrosion and biofouling. The widespread adoption of this technology could drastically alter our energy landscape, providing a consistently reliable renewable energy source.

Tidal barrages, immense structures built across estuaries, trap incoming tidal flows, creating a difference in water level. This difference then drives turbines, generating electricity. The predictable nature of tides makes this a reliable energy source, though the construction of tidal barrages is expensive and complex. Significant environmental impacts, notably alterations to tidal patterns and ecosystems, necessitate careful consideration and mitigation strategies. Ongoing research concentrates on developing more environmentally friendly barrage designs and integrating them effectively within coastal communities. The high capital costs initially limit widespread adoption.

Micro-hydro systems, designed for smaller-scale applications, often utilize the flow of streams or small rivers to generate electricity. These systems are ideal for remote communities or individual properties that lack access to the main grid. They are relatively low-impact environmentally and offer a decentralized, self-sufficient energy solution. Advancements in turbine technology have improved efficiency and reduced costs, making micro-hydro more accessible. The ease of installation and maintenance also contributes to their growing popularity in rural and off-grid settings. While individual output is limited, collectively micro-hydro projects contribute significantly to the broader renewable energy landscape.

Modern hydro projects increasingly prioritize sustainability and environmental responsibility. Life cycle assessments, evaluating the environmental impact from construction to decommissioning, are becoming standard practice. Mitigation strategies aim to minimize disruptions to aquatic ecosystems and address issues such as sediment transport, fish migration, and water quality. Furthermore, integrated approaches combine hydropower with other renewable energy sources, optimizing energy production and reducing reliance on any single technology. These strategies underscore a shift toward a more holistic and sustainable approach to hydropower development. The integration of smart technologies into hydro projects enhances efficiency and monitoring. Remote sensing, advanced data analytics, and predictive modeling assist in optimizing water management, improving turbine performance, and enhancing overall system reliability. Real-time data acquisition and analysis facilitates immediate responses to changes in water flow or system anomalies, leading to optimized energy generation and minimizing potential disruptions. These smart technologies are instrumental in maximizing the efficiency and lifespan of existing and new hydroelectric plants, enhancing their long-term viability. Further improvements are expected as data analytics capabilities evolve and sensors become more sophisticated and resilient.

3.5 Future of Hydropower

Pumped hydro storage, a technology already in use, is poised for significant expansion. Its ability to store excess energy generated from intermittent sources like solar and wind, and then release it during peak demand periods, is invaluable for grid stability. Further research into more efficient pump designs, coupled with advancements in materials science leading to longer-lasting and more durable components, will enhance its overall performance and economic viability. We can expect to see larger-scale pumped hydro projects, possibly integrated with other renewable energy sources, becoming increasingly common.

Advances in micro-hydro technology are opening up new possibilities for smaller-scale energy generation. These systems are ideal for remote communities or areas with limited access to the main power grid. They require less extensive infrastructure than large-scale hydroelectric plants, offering a more sustainable and localized solution. Improvements in turbine design and energy conversion efficiency are continually making micro-hydro a more attractive option, and its integration with smart grids allows for better management and optimization of energy distribution within these smaller systems. This localized approach minimizes environmental impact while providing a reliable energy source.

The development of advanced materials is playing a crucial role in improving the longevity and efficiency of hydropower systems. The use of stronger, more corrosion-resistant materials for dam construction and turbine components reduces maintenance needs and extends the operational lifespan of these facilities. Additionally, the incorporation of smart sensors and predictive maintenance technologies allows for the early detection of potential problems, reducing downtime and optimizing operational efficiency. This proactive approach to maintenance minimizes disruptions and ensures the continued reliability of hydropower as a significant source of electricity.

Harnessing the power of tidal energy is an exciting area of development within the broader hydropower field. Tidal barrages and tidal stream generators offer a predictable and renewable source of energy, particularly in coastal regions with significant tidal ranges. Though the initial investment costs can be substantial, ongoing research into more efficient and cost-effective technologies is paving the way for wider adoption. As our understanding of marine environments improves, the integration of tidal energy projects with minimal environmental impact will become more achievable, unlocking a vast, largely untapped, resource.

Run-of-river hydropower plants are gaining prominence due to their reduced environmental impact compared to traditional dam-based systems. These systems utilize the natural flow of a river to generate electricity, minimizing disruption to river ecosystems and reducing the need for large reservoirs. Design innovations that optimize energy extraction while minimizing environmental impact are actively being pursued. This makes run-of-river systems increasingly attractive, particularly in environmentally sensitive regions where the preservation of natural habitats is paramount. Technological improvements will likely continue to enhance their efficiency and expand their potential.

The integration of hydropower with other renewable energy sources is becoming increasingly important for creating resilient and sustainable energy systems. Hybrid systems that combine hydropower with solar, wind, or geothermal energy can provide a reliable and diverse energy supply, mitigating the intermittency of some renewable sources. Furthermore, the development of advanced energy storage technologies allows for greater flexibility in managing the energy generated by these hybrid systems, ensuring a continuous and stable electricity supply. This approach enhances the overall reliability and sustainability of energy grids worldwide.

A significant challenge for the future of hydropower is addressing its environmental impacts. The construction of large dams can significantly alter river ecosystems, affecting fish migration patterns and water quality. Mitigation strategies are continually being refined, focusing on fish passage solutions, environmental flow requirements, and careful site selection. Life-cycle assessments are increasingly being used to evaluate the environmental impact of hydropower projects throughout their entire lifespan, fostering a more sustainable and environmentally responsible approach to hydropower development. The focus is on minimizing negative consequences while maximizing energy generation.

Addressing climate change necessitates a concerted effort to develop sustainable and resilient energy systems. Hydropower, with its capacity for large-scale energy storage and its relatively low carbon footprint, plays a critical role in this endeavor. Ongoing innovation in technology and design is enhancing the efficiency and environmental performance of hydropower systems, making them an indispensable component of a sustainable energy future. Through responsible planning and implementation, hydropower can contribute significantly to meeting global energy needs while minimizing environmental disruption. Continuous improvements in both technology and sustainability practices will be vital for the future of this critical energy source.

4 Chapter 4: The Nuclear Age

Nuclear fission, the splitting of atomic nuclei, unleashed a new era in power generation. This process releases immense energy, far exceeding that of any previous method. Understanding this fundamental principle is crucial to grasping the potential and challenges of nuclear power. The controlled chain reaction, where the splitting of one nucleus triggers the splitting of others, sustains the energy production in a nuclear reactor. This controlled release is meticulously managed to prevent runaway reactions, a key aspect of nuclear reactor safety. The heart of a nuclear power plant is the reactor itself. Here, uranium fuel rods, enriched to a specific level of U-235, are arranged within a core. This core is submerged in a coolant, usually water, which absorbs the heat generated by fission. The coolant's heat is then used to generate steam, driving turbines connected to generators, much like in conventional steam power plants. This is a simplified explanation, but it highlights the core functionality. However, the complexities extend far beyond this simple process.

Safety features are paramount in nuclear plant design. Multiple layers of containment are incorporated to prevent the release of radioactive materials in case of accidents. These layers, including the reactor vessel itself, various shielding structures, and emergency cooling systems, are engineered to withstand extreme conditions. Regular maintenance and rigorous testing protocols are vital to maintain safety standards. The stringent regulatory frameworks governing nuclear power operations reflect the inherent risks and the need for exceptional safety precautions.

Nuclear waste management remains a significant challenge. Spent nuclear fuel, highly radioactive, requires long-term storage and disposal solutions. This involves both short-term storage at the power plant site and the development of long-term repositories for safe, permanent disposal. Finding suitable geological formations and developing robust containment strategies are key research areas. The long half-lives of many radioactive isotopes necessitates meticulous planning for centuries to come.

The environmental impact of nuclear power is far less than that of fossil fuels regarding greenhouse gas emissions. Nuclear plants do not emit greenhouse gases during operation, contributing to a lower carbon footprint. However, there are potential environmental concerns associated with uranium mining and waste disposal. Careful site selection and strict environmental regulations help to mitigate these impacts. A thorough lifecycle assessment is essential to understand the full environmental effects of nuclear power generation, comparing it with alternatives.

The future of nuclear power is a subject of ongoing debate. There's growing interest in advanced reactor designs, including small modular reactors (SMRs), offering enhanced safety features, increased efficiency, and potentially reduced waste production. These SMRs are designed for modular construction, enabling more flexible deployment and potentially reducing the initial capital investment. This could make nuclear power a more attractive option for diverse energy needs in different geographic locations. However, significant technological advancements and regulatory approvals are still needed for widespread adoption.

Despite the challenges, nuclear power continues to be a significant source of electricity globally. It provides a reliable, baseload power source, independent of weather conditions, unlike solar or wind. This consistent output is vital for grid stability and the reliable supply of electricity to homes and industries. However, the public perception of nuclear power varies widely depending on safety concerns and the management of nuclear waste. Public education and transparency are critical for fostering trust and promoting a rational discussion about the role of nuclear power in a sustainable energy future.

The economics of nuclear power are complex. The high initial capital costs for plant construction are a major factor. However, once operational, nuclear plants have low operating costs and a long lifespan, potentially decades. These long-term considerations must be factored into any economic analysis. The cost competitiveness varies greatly depending on factors like the cost of uranium, construction timelines, and regulatory requirements. The global geopolitical landscape also significantly influences the cost and availability of uranium fuel.

While risks exist, responsible development and stringent regulatory oversight are essential to mitigating them. International cooperation is crucial for sharing best practices in reactor design, safety, and waste management. These global efforts promote improvements and minimize the potential for accidents. Continued research and development in reactor technology and waste management are needed to further enhance the safety and sustainability of nuclear power. This involves exploring advanced reactor designs and developing improved methods for managing nuclear waste.

Nuclear power, with its high energy density and low greenhouse gas emissions, offers a significant contribution to a diversified energy portfolio. However, careful consideration of the challenges and ongoing research efforts are crucial to ensuring its safe and sustainable deployment. Addressing public concerns through open dialogue and transparent communication is paramount to achieving a balanced perspective on the role of nuclear power in meeting global energy demands. Furthermore, international collaboration will be instrumental in advancing safety standards and establishing best practices in nuclear power generation.

4.1 Nuclear Fission Basics

Nuclear fission, at its core, is the process of splitting a heavy atomic nucleus, like uranium-235 or plutonium-239, into lighter nuclei. This splitting releases an immense amount of energy, far exceeding that of chemical reactions. The energy release stems from the conversion of a tiny fraction of the nucleus's mass into energy, as described by Einstein's famous equation, $E=mc^2$. This energy manifests as kinetic energy of the resulting fragments and the emission of neutrons. These neutrons, in turn, can initiate further fission reactions in surrounding nuclei, leading to a chain reaction – a self-sustaining process critical for sustained power generation.

Understanding the chain reaction is paramount. A critical mass of fissile material is necessary. This is the minimum amount of material required for the chain reaction to sustain itself. If the mass is too small, many neutrons escape without causing further fissions, and the reaction fizzles out. Conversely, a supercritical mass leads to an uncontrolled, exponentially increasing chain reaction, as seen in nuclear weapons. Reactor design focuses on maintaining a carefully controlled, subcritical or slightly supercritical state to regulate power output.

Control of the chain reaction relies heavily on neutron moderators and control rods. Moderators, often water or graphite, slow down the fast neutrons produced during fission. Slower neutrons are more effective at causing further fissions in uranium-235. Control rods, typically made of neutron-absorbing materials like cadmium or boron, are inserted into the reactor core to absorb neutrons and regulate the reaction rate. By adjusting the position of the control rods, the operator can precisely control the rate of fission and, consequently, the power output of the reactor.

The energy released during fission appears primarily as heat. This heat is transferred to a coolant, typically water, which circulates through the reactor core. The heated coolant then transfers its thermal energy to a secondary loop, which generates steam to drive turbines connected to electric generators. This is a fundamental principle behind the operation of most nuclear power plants, transforming the raw energy of atomic fission into usable electricity. Understanding this heat transfer process, including pressure, temperature, and flow rates, is vital for safe and efficient operation. Specific designs vary based on factors such as reactor type (e.g., pressurized water reactor, boiling water reactor, CANDU reactor) and safety features. Reactor safety is paramount. Multiple layers of safety systems are incorporated into nuclear power plant designs to prevent accidents and mitigate potential consequences. These systems include redundant safety systems, emergency core cooling systems, and containment structures designed to prevent the release of radioactive materials into the environment. Regular inspections, stringent regulations, and operator training are critical to maintaining the safe operation of nuclear power plants. Furthermore, managing nuclear waste, the radioactive byproduct of fission, requires careful planning and long-term solutions. The spent fuel remains radioactive for thousands of years, requiring secure storage and ultimately, ideally, processing and disposal.

While nuclear fission offers a powerful, high-energy density means of electricity generation, it does present challenges. The potential for accidents, albeit mitigated by robust safety systems, remains a concern. The long-term storage and management of nuclear waste pose significant logistical and environmental hurdles. Public perception and acceptance of nuclear power are also influenced by these factors. However, the continued development of advanced reactor designs and enhanced safety features alongside careful waste management strategies aims to address these challenges and enhance the safety and sustainability of nuclear power as a significant contributor to the global energy mix. A complete picture requires a balanced understanding of the scientific principles, engineering challenges and societal impacts surrounding nuclear fission. Further research, advancements in technology, and informed public discourse all play a role in the continued evolution of nuclear energy.

4.2 Nuclear Power Plant Design

Nuclear power plants are complex systems demanding meticulous design to ensure safe and efficient energy production. The fundamental principle, nuclear fission, relies on splitting heavy atoms like uranium-235, releasing immense energy in the form of heat. This heat is then harnessed to generate electricity. A crucial component is the reactor core, housing the nuclear fuel and control rods. These rods, made of neutron-absorbing materials like boron or cadmium, regulate the fission rate, controlling the power output and preventing runaway reactions. Surrounding the core is a coolant system, typically water or a gas, which absorbs the heat generated by fission.

The design of the reactor itself varies. Pressurized water reactors (PWRs) are the most common type, using high-pressure water as both coolant and moderator, slowing down neutrons to increase fission efficiency. Boiling water reactors (BWRs) use boiling water as the coolant, creating steam directly within the reactor. Both designs prioritize safety features including multiple containment structures to prevent the release of radioactive materials in case of an accident. This layered approach involves robust steel vessels, concrete structures, and emergency cooling systems. Detailed calculations and simulations are essential during the design process to account for thermal stresses, pressure fluctuations, and other factors affecting plant integrity.

Beyond the reactor, the plant incorporates a steam turbine system. The superheated coolant, either steam directly from the BWR or steam generated from the PWR's heated water in a secondary loop, drives the turbines, converting thermal energy into mechanical energy. These turbines are connected to electrical generators, producing electricity which is then stepped up in voltage by transformers and fed into the power grid. This entire process, from nuclear fission to electricity generation, involves sophisticated instrumentation and control systems to continuously monitor and regulate parameters like temperature, pressure, and radiation levels. This includes real-time monitoring, automatic safety shutdowns, and human intervention capabilities, all critical for ensuring safe operation.

Efficient waste management is another crucial aspect of nuclear power plant design. Spent nuclear fuel, highly radioactive, needs careful handling and storage. On-site storage pools are commonly used for temporary storage, while long-term solutions like geological repositories are still under development. The design needs to incorporate these stages from the beginning. The entire lifecycle, from fuel fabrication to decommissioning after the plant's operational life, needs careful planning. The choice of materials for components is also critical, considering resistance to corrosion, radiation damage, and high temperatures. The selection involves rigorous testing and evaluation to guarantee long-term performance and safety. Every aspect, from the reactor core's materials to the protective barriers and waste handling strategies, involves intricate engineering design.

Furthermore, effective safety systems are paramount. Multiple redundant layers are employed to prevent accidents, including emergency core cooling systems (ECCS), containment systems, and multiple independent safety instrumentation channels. These systems are designed to automatically shut down the reactor in case of abnormal conditions, preventing core damage and radioactive releases. Regular inspections, maintenance, and operator training are crucial in ensuring the long-term safety and reliability of the plant. The design also has to consider the plant's location and potential environmental impacts, including seismic activity, flooding, and extreme weather events. Risk assessments and mitigation strategies are integrated into the design process to minimize the probability and consequences of potential hazards. Even security measures against potential sabotage or terrorism are incorporated into modern plant designs. Nuclear power plant design is a multifaceted engineering endeavor, integrating expertise from various disciplines to deliver safe, efficient, and reliable electricity generation.

5 Chapter 5: Fossil Fuel Power

Coal-fired power plants represent a mature technology, relying on the combustion of coal to heat water, creating steam that drives turbines connected to generators. These plants are typically large-scale facilities, demanding significant upfront investment but capable of providing substantial baseload power—a consistent supply of electricity to meet constant demand. The process involves complex systems for coal handling, combustion, steam generation, and emission control, underscoring the technological sophistication required for efficient and safe operation. Modern coal plants incorporate advanced technologies to enhance efficiency and minimize emissions, although significant environmental challenges remain.

Natural gas power plants, in contrast, offer a faster response time and greater flexibility than coal plants. Natural gas combustion generates heat for steam turbines or drives gas turbines directly, offering a more agile power source suitable for peak demand periods. These plants are often smaller and less expensive to build than coal plants, making them a more attractive option for smaller-scale electricity generation or to supplement baseload power from other sources. The relative cleanliness of natural gas combustion compared to coal is a key advantage, although the extraction and transportation of natural gas can also have environmental consequences.

The environmental impact of fossil fuel power generation is a significant concern. Coal combustion releases substantial quantities of carbon dioxide (CO2), a major greenhouse gas contributing to climate change. Furthermore, coal plants emit sulfur dioxide (SO2) and nitrogen oxides (NOx), leading to acid rain and respiratory problems. While natural gas combustion produces fewer emissions than coal, it still contributes to CO2 emissions and can release methane, a potent greenhouse gas, during extraction and transportation. Stricter emission regulations and the development of carbon capture and storage (CCS) technologies aim to mitigate these environmental impacts, although the effectiveness and cost-effectiveness of these measures remain subjects of ongoing debate and research. The long-term sustainability of fossil fuel power is increasingly questioned, prompting a global shift towards cleaner and more sustainable energy sources. This transition necessitates strategic planning, technological innovation, and robust regulatory frameworks to ensure a reliable and environmentally responsible energy future. Balancing energy needs with environmental protection requires careful consideration of various factors, including energy security, economic viability, and the environmental consequences of different energy sources. The transition to cleaner energy sources is a complex process requiring careful management and collaboration among governments, industries, and researchers.

The complexities of fossil fuel power extend beyond mere combustion and emission control. The entire lifecycle, from extraction and transportation to plant operation and decommissioning, presents environmental and social challenges. Mining for coal involves habitat destruction and potential for water and air pollution. Similarly, natural gas extraction, particularly through fracking, raises concerns about groundwater contamination and induced seismicity. These considerations highlight the need for comprehensive life-cycle assessments of fossil fuel power plants, taking into account environmental and social impacts throughout the entire process. Moreover, the geopolitical implications of fossil fuel dependence cannot be overlooked, as reliance on these resources can create vulnerabilities to price volatility and international conflicts. Diversifying energy sources and fostering energy independence are therefore crucial steps towards enhancing

energy security and reducing geopolitical risks.
Addressing the environmental challenges associated with fossil fuel power is paramount. The development and deployment of CCS technologies offer a potential pathway to reducing CO2 emissions from these plants, although widespread adoption faces technological, economic, and logistical hurdles. Furthermore, improving plant efficiency and optimizing operational parameters can help minimize emissions and improve resource utilization. The integration of fossil fuel plants into broader energy systems, considering their role alongside renewable energy sources, presents opportunities for enhanced grid flexibility and reduced reliance on fossil fuels. This integration necessitates advanced grid management systems capable of accommodating intermittent renewable energy sources while maintaining grid stability and reliability. The future of fossil fuel power lies in managing its transition to a reduced role in energy production, emphasizing efficiency, emissions reduction, and responsible environmental stewardship. This involves fostering innovation in CCS technology, promoting energy efficiency measures, and exploring strategies for integrating fossil fuel plants into a broader energy mix that increasingly emphasizes renewable resources. Ultimately, the responsible use of fossil fuels, accompanied by a focused transition towards cleaner energy options, is essential for a sustainable energy future.

5.1 Coal-fired Power Plants

Coal-fired power plants represent a significant, albeit increasingly controversial, chapter in the history of electricity generation. These plants, for decades the workhorses of national grids worldwide, rely on the combustion of coal to produce steam, which then drives turbines connected to generators. This process, while relatively straightforward in principle, involves complex engineering and numerous stages. From the initial mining and transportation of coal, to the intricate systems managing combustion, steam generation, and waste disposal, the entire operation demands significant technical expertise and careful monitoring.

The heart of a coal-fired power plant is the boiler, a massive pressure vessel where coal is burned at extremely high temperatures. This combustion process generates heat that converts water into high-pressure steam. The steam then flows to the turbine, a sophisticated piece of machinery with intricately designed blades. The high-velocity steam spins the turbine blades, transferring its kinetic energy to a rotating shaft connected to the generator. This generator, utilizing principles of electromagnetism, converts the mechanical energy into electrical energy, ready for transmission to the power grid. However, the seemingly simple transformation of coal into electricity masks a complex web of technological challenges. Maintaining optimal combustion conditions, ensuring consistent steam pressure, and managing the potentially corrosive effects of high-temperature steam on plant components are just some of the engineering hurdles. Furthermore, efficient heat transfer is critical to maximizing the plant's overall efficiency, minimizing fuel consumption, and reducing operational costs. Sophisticated control systems, constantly monitoring and adjusting various parameters, are vital for safe and efficient operation.

Beyond the technical aspects, coal-fired plants present substantial environmental challenges. The combustion of coal releases significant quantities of greenhouse gases, primarily carbon dioxide, a major contributor to climate change. Additionally, the combustion process generates various pollutants, including sulfur oxides and nitrogen oxides, contributing to acid rain and respiratory problems. Furthermore, coal mining itself can have severe environmental consequences, such as habitat destruction and water contamination. Therefore, considerable efforts are dedicated to implementing pollution control technologies, such as scrubbers to remove sulfur oxides and electrostatic precipitators to remove particulate matter.

Modern coal-fired power plants incorporate advanced technologies designed to improve efficiency and mitigate environmental impacts. Supercritical and ultra-supercritical boilers operate at higher pressures and temperatures, achieving greater efficiency in converting heat energy to electrical energy. Integrated gasification combined cycle (IGCC) plants offer a more environmentally friendly approach by gasifying coal before combustion, reducing emissions. However, these advanced technologies come at a higher initial investment cost. The future of coal-fired power plants is uncertain, largely due to the escalating concerns surrounding climate change and air pollution. Many countries are phasing out coal-fired power generation, shifting towards cleaner and more sustainable energy sources. Nevertheless, coal remains a significant energy source in many parts of the world, particularly in regions with abundant coal reserves and limited access to alternative energy technologies. The challenge lies in finding a balance between meeting the world's energy demands and minimizing the environmental impact of coal-fired power plants, perhaps through carbon capture and storage technologies or a gradual transition to cleaner energy alternatives.

Considering the complex interplay between energy needs, economic factors, and environmental concerns, the operational life of existing coal plants and the construction of new ones are subject to rigorous scrutiny. Governments and regulatory bodies around the globe are actively involved in establishing stringent emission standards and promoting cleaner energy solutions. This shift reflects the evolving understanding of climate change and its implications for future energy strategies. As such, future generations of power engineers will undoubtedly focus on the development of increasingly efficient and sustainable electricity generation technologies, gradually phasing out the dependence on fossil fuels. The legacy of coal-fired power plants, though significant in powering industrial growth, is now inextricably linked to the global transition towards a decarbonized energy future. This transition will require both technological innovation and substantial policy changes to ensure a secure and sustainable energy supply for all. The development and deployment of effective energy storage technologies will also play a crucial role in supporting the intermittent nature of many renewable sources. Ultimately, the successful transition requires collaborative efforts among governments, industries, and researchers, driving the innovation and deployment of sustainable energy alternatives while managing the environmental consequences of existing infrastructure. This shift represents a profound technological and societal challenge, demanding creative solutions and a global commitment to a cleaner energy future for all.

5.2 Natural Gas Power Plants

Natural gas, a hydrocarbon mixture primarily composed of methane, has become a cornerstone of modern electricity generation. Its widespread availability, relatively clean combustion compared to coal, and adaptability to various power plant designs have contributed to its significant role in the global energy mix. Unlike coal, natural gas combustion produces fewer pollutants, making it a transitional fuel source in the shift towards renewable energy. This cleaner profile, however, doesn't eliminate environmental concerns entirely. Greenhouse gas emissions, though lower than coal, still pose a significant challenge.

The fundamental principle behind natural gas power generation is straightforward: burning natural gas to heat water, producing high-pressure steam that drives turbines connected to generators. This process, a refined version of earlier steam power methods, leverages sophisticated technology to maximize efficiency and minimize environmental impact. Modern natural gas plants boast advanced combustion systems designed to reduce nitrogen oxide (NOx) emissions, a significant air pollutant. Furthermore, the implementation of Selective Catalytic Reduction (SCR) technology further minimizes NOx levels, enhancing the overall environmental performance. The integration of these technological advancements underscores the ongoing efforts to mitigate the environmental footprint of fossil fuel-based electricity generation.

These power plants aren't simply large-scale versions of earlier designs. They incorporate sophisticated controls and monitoring systems allowing for precise management of combustion and energy output. This real-time control optimizes efficiency and responsiveness to fluctuating energy demands on the grid. The integration of advanced digital controls contributes to grid stability and reliability, ensuring a consistent power supply. This intelligent management system significantly improves the overall operational efficiency, leading to cost savings and reduced environmental impact. Operators can fine-tune the power output to match real-time demand, minimizing energy waste and maximizing the plant's contribution to the energy grid.

A significant aspect of natural gas power plant design is the turbine technology employed. Advanced gas turbines, often featuring multiple stages of compression and expansion, achieve exceptionally high thermal efficiencies. These turbines, significantly more efficient than their earlier counterparts, translate directly into reduced fuel consumption and lower operating costs. This technological advancement highlights the continuous drive for efficiency improvements within the power generation sector. Furthermore, the development of combined-cycle power plants further enhances efficiency. These plants incorporate both gas turbines and steam turbines, using the waste heat from the gas turbine to generate additional power through a steam cycle.

However, the future of natural gas in electricity generation is not without its complexities. The growing emphasis on decarbonization and the transition towards renewable energy sources pose a significant challenge. While natural gas is a comparatively cleaner fossil fuel, its contribution to greenhouse gas emissions cannot be ignored. The ongoing development and deployment of carbon capture and storage (CCS) technology aim to address this concern by capturing CO2 emissions before they enter the atmosphere. Successful implementation of CCS could significantly extend the operational lifespan of natural gas plants while mitigating their environmental impact. This technology, however, is still in its developmental stages and faces technological and economic hurdles that must be overcome.

The economics of natural gas power plants are complex and dynamic. Natural gas prices fluctuate based on global supply and demand, influencing the overall cost-effectiveness of electricity generation. The construction costs of these plants can be substantial, requiring significant financial investment. Nevertheless, the comparatively lower operating costs compared to other fossil fuel sources, coupled with the relative speed of construction compared to renewable energy projects like large-scale solar or wind farms, makes them an attractive option, especially for bridging the transition towards a fully renewable energy system. Policy incentives and government regulations further influence the economic landscape, promoting or hindering the development of natural gas power plants.

The environmental impact of natural gas power plants, while less severe than coal-fired plants, remains a significant concern. Methane leakage throughout the natural gas supply chain, from extraction to distribution, contributes to greenhouse gas emissions. This leakage significantly impacts the overall carbon footprint, making it crucial to minimize methane emissions across the entire supply chain. Stricter regulations and improved infrastructure are necessary to mitigate this environmental concern. Furthermore, the disposal of the residual ash from natural gas combustion requires careful management to prevent environmental contamination.

The design and construction of natural gas power plants are complex engineering endeavors, necessitating specialized expertise and rigorous safety protocols. The high temperatures and pressures within these plants necessitate robust materials and sophisticated control systems to ensure safe and reliable operation. Regular maintenance and inspection are paramount to prevent equipment failure and ensure the safety of plant personnel and the surrounding community. This emphasis on safety underscores the critical role of proper engineering design and meticulous operation in the successful management of natural gas power plants. Ultimately, responsible operation and ongoing technological advancement will determine the long-term viability of natural gas in the evolving energy landscape.

5.3 Environmental Concerns

The reliance on fossil fuels for electricity generation presents a significant environmental challenge. Coal and natural gas power plants, while efficient in producing large quantities of electricity, release substantial amounts of greenhouse gases like carbon dioxide, a primary contributor to climate change. These emissions impact air quality, leading to respiratory illnesses and other health problems, not just for those living near power plants but across wide geographical areas due to atmospheric dispersion. Furthermore, the extraction and transportation of fossil fuels often result in habitat destruction and pollution of water sources, harming ecosystems and wildlife.

Beyond greenhouse gases, the combustion of fossil fuels releases other pollutants, including sulfur dioxide and nitrogen oxides. These contribute to acid rain, damaging forests and aquatic life. The particulate matter released can also negatively affect human health, exacerbating existing cardiovascular and respiratory conditions. Addressing these environmental consequences requires a multifaceted approach involving both technological improvements and policy changes.

Technological advancements offer some potential solutions. Carbon capture and storage (CCS) technologies aim to capture CO2 emissions from power plants before they enter the atmosphere, storing them underground. While promising, CCS technology is still expensive and faces challenges in terms of scalability and long-term storage security. Improved combustion techniques can also reduce emissions by maximizing energy efficiency and minimizing the formation of pollutants. However, even with these improvements, the inherent environmental impact of burning fossil fuels remains a concern.

Policy interventions play a crucial role in mitigating environmental damage. Regulations limiting greenhouse gas emissions, such as carbon pricing mechanisms or emission trading schemes, incentivize power plants to adopt cleaner technologies and reduce their carbon footprint. Subsidies and tax breaks for renewable energy sources make them more economically competitive compared to fossil fuels, driving the transition towards a sustainable energy mix. Stringent environmental regulations regarding air and water quality also help reduce the adverse effects of fossil fuel power generation. These policies must be implemented effectively and consistently across nations to maximize their global impact.

Investing in research and development for cleaner energy alternatives is critical. While renewable energy sources such as solar, wind, and geothermal power are growing rapidly, continuous improvements in efficiency and cost-effectiveness are needed to ensure their widespread adoption. Developing advanced energy storage technologies is also essential to overcome the intermittency of renewable energy sources, which means their power output may vary depending on weather conditions. Such advancements are critical to ensuring a reliable and sustainable power supply. Without continued innovation and development, the transition to cleaner energy sources will be significantly slower and more challenging.

Moreover, considering the entire lifecycle of power plants—from extraction and transportation of fuels to construction, operation, and decommissioning—is vital for a holistic assessment of environmental impacts. For example, coal mining causes considerable land disturbance and potential for water contamination, which must be taken into account when evaluating the overall environmental footprint of coal-fired power generation. Similarly, the disposal of waste products from power plants needs careful planning and management to prevent environmental damage. A comprehensive approach is needed, one that recognizes the interconnectedness of various environmental aspects and promotes a sustainable and responsible energy future. This will require strong policy support, technological innovation, and a fundamental shift in how we approach power generation. Ultimately, the transition away from fossil fuels is not merely an environmental imperative; it's also a crucial step towards ensuring a healthier and more sustainable future for all. The urgency of this transition cannot be overstated.

6 Chapter 6: Renewable Energy Sources

Harnessing the sun's energy through photovoltaic cells, solar power offers a clean and sustainable alternative. These cells, made of semiconductor materials like silicon, convert sunlight directly into electricity. Technological advancements have significantly improved their efficiency, reducing production costs and making solar power increasingly competitive with traditional energy sources. Large-scale solar farms, often located in sun-rich regions, generate significant amounts of electricity, feeding into the national grid and powering entire communities. However, the intermittent nature of solar energy, dependent on weather conditions, necessitates effective energy storage solutions, a topic we'll explore further in Chapter 7. Furthermore, the environmental impact of manufacturing solar panels and their eventual disposal requires careful consideration for a truly sustainable energy future.

Wind energy harnesses the kinetic energy of moving air using wind turbines. These turbines, comprising large blades mounted on a tower, convert wind's force into rotational energy, which is then converted to electricity via a generator. Advances in turbine design, including longer blades and optimized aerodynamics, have increased their energy capture efficiency. Wind farms, collections of numerous turbines, are now capable of generating substantial quantities of power, sometimes even exceeding the output of traditional power plants. Strategic placement of wind farms, considering wind patterns and minimizing environmental impact, is crucial for maximizing efficiency and minimizing disruption to ecosystems. The intermittency of wind, similar to solar power, presents challenges, further highlighting the importance of grid modernization and energy storage.

Geothermal energy taps into the Earth's internal heat. This heat, originating from radioactive decay deep within the planet, can be harnessed to generate electricity using geothermal power plants. These plants use steam or hot water extracted from underground reservoirs to drive turbines and generate electricity. Geothermal energy offers a reliable, baseload power source, unlike solar and wind. However, geothermal resources are geographically limited to areas with active volcanic activity or significant geothermal gradients. Furthermore, the potential for induced seismicity and the environmental impact of geothermal fluid extraction and disposal necessitate careful planning and monitoring.

Biomass energy utilizes organic matter, such as wood, agricultural residues, and dedicated energy crops, as a fuel source. Biomass can be burned directly in power plants or converted into biofuels, which can be used in internal combustion engines or power turbines. Advances in biomass conversion technologies have improved efficiency and reduced emissions. However, sustainability concerns remain, primarily regarding land use and deforestation, if biomass resources aren't managed responsibly. Efficient and sustainable biomass production, along with advanced processing technologies, is vital to minimizing negative environmental impacts and maximizing the contribution of biomass to a sustainable energy mix.

Ocean energy, encompassing various technologies, harnesses the power of ocean currents, waves, and tidal movements. Ocean thermal energy conversion (OTEC) utilizes the temperature difference between warm surface water and cold deep water to generate electricity. Wave energy converters capture the kinetic energy of waves to produce electricity, while tidal barrages utilize the rise and fall of tides to drive turbines. These technologies are still relatively underdeveloped compared to solar, wind, and geothermal, but they hold substantial potential, particularly in coastal regions with significant ocean energy resources. Technological advancements, focused on improving efficiency, durability, and cost-effectiveness, are essential to unlock the full potential of ocean energy. Additionally, the environmental impacts of these technologies, including potential effects on marine life and ecosystems, must be carefully assessed and mitigated.

6.1 Solar Power Technology

Solar power, harnessing the sun's energy, has emerged as a vital component of renewable energy generation. Its technology, constantly evolving, offers diverse options for electricity production, from small-scale residential systems to massive utility-scale solar farms. Photovoltaic (PV) cells, the heart of solar power, convert sunlight directly into electricity through the photovoltaic effect. This effect, a fascinating interplay of photons and semiconductor materials, creates a flow of electrons, generating a direct current (DC) electricity that can then be inverted to alternating current (AC) for use in homes and businesses. The efficiency of PV cells, a crucial factor determining the overall effectiveness of a solar power system, has steadily increased over the years, driven by advancements in materials science and manufacturing techniques. Higher efficiency translates directly into greater energy output per unit area, ultimately leading to cost reductions and improved performance. Silicon, the most common material used in PV cells, undergoes rigorous refinement processes to enhance its crystalline structure and improve its light-to-electricity conversion capabilities.

Different types of PV cells exist, each with unique characteristics and applications. Monocrystalline silicon cells, known for their high efficiency and black appearance, are often preferred for residential and commercial installations where space is at a premium. Polycrystalline silicon cells, distinguished by their slightly lower efficiency and characteristic blueish hue, offer a more cost-effective alternative, making them suitable for large-scale projects. Thin-film solar cells, a more recent development, utilize materials like cadmium telluride or copper indium gallium selenide (CIGS) deposited as thin layers on a substrate. These cells offer advantages in terms of flexibility and manufacturing costs, although they typically have lower efficiencies than crystalline silicon cells. The choice of PV cell type depends on several factors, including the available space, budget constraints, and specific energy needs. Furthermore, advancements in concentrating solar power (CSP) technology offer another avenue for harnessing solar energy. CSP systems utilize mirrors or lenses to concentrate sunlight onto a receiver, generating heat that is used to drive a heat engine and produce electricity. This technology, while requiring significant land area, can achieve higher efficiency compared to PV systems, particularly in regions with abundant sunshine.

The integration of solar power systems involves several key components beyond the PV cells themselves. Solar panels, composed of multiple PV cells interconnected in series and parallel, are typically mounted on rooftops, ground mounts, or specialized tracking systems that follow the sun's movement throughout the day. Inverters are essential devices that convert the DC electricity produced by the panels into AC electricity compatible with the electrical grid. A charge controller regulates the flow of electricity from the solar panels to batteries, protecting the batteries from overcharging and ensuring optimal performance. This is especially important in off-grid systems or those incorporating battery storage. Battery banks, used in off-grid or backup power systems, store excess energy generated during the day for use during periods of low solar irradiance or at night. The choice of battery type and capacity depends on factors such as energy demand, system size, and budget considerations. Finally, wiring, connectors, and mounting hardware are crucial elements that ensure the safe and efficient operation of the entire solar power system. These elements must be carefully chosen to meet the relevant safety standards and regulations.

Beyond the technical aspects, the economic and environmental benefits of solar power are compelling reasons for its growing adoption. The decreasing cost of solar panels has made solar energy increasingly affordable, particularly when combined with government incentives and subsidies. Moreover, the environmental advantages are substantial. Solar power produces no greenhouse gas emissions during operation, making it a clean and sustainable energy source. This aligns perfectly with global efforts to mitigate climate change and reduce reliance on fossil fuels. The environmental impact during manufacturing is increasingly considered, with continuous research into more sustainable materials and manufacturing processes. However, challenges remain, including land-use requirements for large-scale solar farms, and the intermittent nature of solar energy. Energy storage technologies, such as advanced battery systems and pumped hydro storage, are crucial for addressing the intermittency issue and ensuring a reliable and consistent energy supply. This continues to drive significant innovation and investment in the field. Further research and development are focused on improving the efficiency of solar cells, reducing manufacturing costs, and developing more efficient energy storage solutions, paving the way for the widespread adoption of solar power as a key component of a sustainable energy future. The potential of solar energy to power our world sustainably is enormous, requiring continuous exploration and advancement in its technology and implementation. The future of solar power depends on the convergence of technological innovation, policy support, and a global commitment to sustainability.

6.2 Wind Energy Farms

Wind energy farms represent a significant step towards a sustainable energy future, converting a naturally occurring resource into usable electricity. Their design and operation involve intricate considerations, from site selection and turbine technology to grid integration and environmental impact assessments. Understanding these complexities is crucial for anyone aiming to contribute to the renewable energy sector.

The initial phase of establishing a wind farm necessitates meticulous site analysis. Factors such as average wind speed, wind direction consistency, and turbulence intensity are rigorously evaluated using sophisticated meteorological data and advanced modeling techniques.

Furthermore, assessing the local topography, proximity to transmission lines, and potential environmental impacts – including effects on bird and bat populations – are paramount. These considerations directly influence the farm's overall efficiency and long-term viability.
Turbine selection is another critical aspect. Modern wind turbines are marvels of engineering, featuring advanced blade designs optimized for aerodynamic efficiency and durable materials designed to withstand harsh weather conditions. The choice of turbine type—horizontal-axis or vertical-axis—depends on several site-specific parameters. Horizontal-axis wind turbines, the most common type, are highly efficient in consistent wind conditions, while vertical-axis turbines offer advantages in areas with variable wind direction. Furthermore, ongoing technological advancements continuously improve turbine performance, increasing energy capture and reducing maintenance requirements.
Once turbines are selected and installed, their integration into the existing power grid is a critical process. This involves the careful planning and construction of transmission lines and substations to efficiently transport the generated electricity to consumers. The intermittency of wind power necessitates sophisticated grid management strategies to ensure grid stability and prevent power outages. This often involves incorporating energy storage systems, such as batteries or pumped hydro storage, to compensate for fluctuations in wind availability. These storage solutions are vital for maximizing the utilization of wind energy.
Beyond the technical aspects, the environmental impact of wind farms deserves close attention. While wind energy is considered a clean energy source, potential impacts on wildlife, particularly birds and bats, must be thoroughly assessed and mitigated through careful site selection and operational practices. Furthermore, the visual impact on landscapes is a key concern that requires thoughtful consideration, balancing energy production with community acceptance and aesthetic considerations. Environmental studies form an integral part of the planning process for wind farms, guiding responsible development and minimizing negative effects.

The economic aspects of wind energy farms are also significant. The initial investment for establishing a wind farm is substantial, requiring careful financial planning and securing appropriate funding. However, the long-term operational costs are relatively low, and the increasing affordability of wind energy technology makes it increasingly competitive with traditional power generation sources. Furthermore, wind farms often create local jobs, contributing to economic growth in the surrounding communities. Financial models forecasting profitability are essential for developers to consider before embarking on such substantial projects.
Ultimately, wind energy farms represent a dynamic and rapidly evolving field, demanding a multidisciplinary approach encompassing engineering, meteorology, environmental science, and economics. Future advancements will likely focus on improving turbine technology, optimizing grid integration techniques, and developing more efficient energy storage solutions. Furthermore, the increasing integration of smart grid technologies will further enhance the effectiveness of wind energy farms, ensuring their contribution to a sustainable and reliable electricity supply. This multifaceted process necessitates interdisciplinary teamwork and continuous innovation, driving the continued growth and sophistication of this vital renewable energy source. The field is therefore continuously attracting research and development efforts, pushing technological boundaries and enhancing efficiency, cost-effectiveness, and environmental compatibility.

6.3 Geothermal Energy

Geothermal energy taps into the Earth's internal heat, a virtually inexhaustible resource. Unlike solar or wind power, geothermal energy provides a consistent, baseload power source, unaffected by weather patterns. This reliability is a significant advantage, contributing to grid stability and reducing reliance on intermittent renewables. Harnessing this energy involves drilling deep wells to access reservoirs of hot water or steam, which are then used to drive turbines and generate electricity. The technology, while requiring substantial upfront investment, offers long-term operational benefits and environmental advantages compared to fossil fuels.

The process begins with identifying suitable geothermal resources, often through geological surveys and geophysical techniques. These surveys pinpoint areas with high geothermal gradients, indicating the presence of significant underground heat. Once a suitable location is identified, exploratory drilling is undertaken to assess the temperature, pressure, and flow rate of the geothermal reservoir. This crucial step determines the feasibility and economic viability of a geothermal power plant. Careful site selection minimizes environmental impact and maximizes energy extraction efficiency. Rigorous environmental impact assessments must be performed to evaluate potential effects on groundwater, surface water, and local ecosystems.

Different types of geothermal power plants exist, each suited to specific geological conditions. Enhanced Geothermal Systems (EGS) are particularly promising, allowing the exploitation of geothermal resources in areas lacking naturally occurring hot water reservoirs. EGS involves creating artificial reservoirs by fracturing hot rock deep underground, circulating water to extract heat and then using this heated water to generate electricity. This technology expands the potential of geothermal energy significantly, opening up vast new areas for development. While EGS is still under development, ongoing research and technological advancements are continually improving its efficiency and cost-effectiveness.

The environmental impact of geothermal energy is significantly less than that of fossil fuels. Geothermal power plants produce minimal greenhouse gas emissions, contributing to a cleaner energy future. Moreover, the land footprint of geothermal plants is relatively small compared to other renewable energy sources like solar or wind farms. However, some potential environmental concerns remain. These include induced seismicity, the possibility of groundwater contamination, and the release of trace amounts of greenhouse gases from geothermal fluids. Mitigation strategies, such as careful well design and injection practices, aim to minimize these potential risks, maximizing the environmental benefits of geothermal energy.

Operating a geothermal power plant involves ongoing monitoring and maintenance of the wells and power generation equipment. Regular inspection of the wells is crucial to detect any potential issues and prevent problems like scaling or corrosion. Furthermore, careful management of the geothermal fluids is essential to prevent environmental contamination. Effective monitoring and maintenance programs ensure efficient operation and extended lifespan of geothermal power plants. The long-term economic viability of geothermal energy is heavily reliant on these operational aspects, highlighting the need for sustained technological innovation and expertise. The overall economic viability is influenced by a variety of factors, including resource availability, plant size, and the prevailing electricity prices. Therefore, economic analysis is critical for project planning and investment decisions. Careful consideration of various economic parameters is crucial for achieving the project's financial success.
The future of geothermal energy looks bright. Technological advancements are continuously enhancing the efficiency and cost-effectiveness of geothermal power generation. EGS technology, in particular, holds immense potential for expanding geothermal energy production. Furthermore, combined heat and power (CHP) systems can utilize the heat extracted from geothermal resources for various applications such as heating and industrial processes, significantly improving overall energy efficiency. Research and development efforts focusing on enhanced reservoir management, improved drilling techniques, and advanced heat exchangers will further optimize geothermal energy extraction and minimize costs. International collaboration and technology transfer are vital in promoting the global adoption of geothermal energy, helping to address climate change and improve energy security. Therefore, investment in research and development is a crucial factor to ensure the sustainable growth of this energy sector. As the world transitions towards cleaner energy sources, geothermal energy is poised to play a significant role in meeting global energy demands sustainably.

6.4 Biomass Energy

Biomass energy harnesses the chemical energy stored within organic matter, converting it into usable electricity. This organic matter encompasses a wide range of sources, from agricultural residues like corn stalks and rice husks, to dedicated energy crops such as switchgrass and willow trees, and even municipal solid waste. The process, unlike the direct combustion often associated with biomass, typically involves sophisticated technologies to maximize energy efficiency and minimize environmental impact.

Several methods exist for converting biomass into electricity. Gasification, a thermochemical process, converts biomass into a syngas (synthesis gas) composed primarily of hydrogen and carbon monoxide. This syngas can then be used to fuel gas turbines or be further processed into biofuels, offering a flexible pathway for electricity generation. Pyrolysis, another thermochemical process, uses heat to break down biomass in the absence of oxygen, yielding biochar, bio-oil, and syngas. Each product has potential energy applications; bio-oil, for example, can be refined into transportation fuels or used directly in combustion engines for electricity generation.

Anaerobic digestion, a biological process, utilizes microorganisms to break down organic matter in the absence of oxygen, producing biogas, a mixture primarily of methane and carbon dioxide. This biogas can be directly combusted in engines to generate electricity, or upgraded to biomethane, which can be injected into natural gas pipelines, integrating biomass energy into existing infrastructure. The digestate, a byproduct of anaerobic digestion, is a valuable fertilizer, creating an opportunity for circular economy practices within agriculture. The efficiency of each process varies depending on several factors, including the type of biomass used, the specific technology employed, and the overall system design. Careful optimization of these factors is critical for maximizing energy output.

Choosing the optimal biomass conversion technology depends on several key considerations. The type and availability of biomass feedstock strongly influence the selection process. Agricultural residues, for instance, might lend themselves to gasification or anaerobic digestion, while dedicated energy crops could be more suitable for pyrolysis. The scale of operation also plays a significant role; smaller-scale operations might favor simpler technologies like direct combustion or anaerobic digestion, while large-scale power generation might benefit from the higher efficiencies of gasification or advanced pyrolysis systems. Economic factors, including the cost of biomass feedstock, the capital investment in technology, and the operational costs, all weigh heavily in the decision-making process. Environmental considerations, such as greenhouse gas emissions and land-use changes, are paramount in responsible biomass energy development.

Environmental impacts are a critical aspect of biomass energy production. While biomass is considered a carbon-neutral energy source, responsible sourcing and processing are essential to mitigate potential negative environmental effects. Sustainable forestry practices, for example, are vital to ensure that biomass harvesting does not lead to deforestation or habitat loss. The efficient and complete combustion of biomass is crucial to minimizing air pollution, while proper management of waste products, including ash and digestate, is necessary to prevent soil and water contamination. Life-cycle assessments, evaluating the environmental impact throughout the entire production process, from feedstock cultivation to electricity generation and waste disposal, are essential tools for making informed decisions. These assessments help identify areas where environmental improvements can be achieved, driving the development of more environmentally benign biomass energy systems.

The future of biomass energy lies in innovation and sustainability. Research and development efforts are focused on improving the efficiency of existing technologies and developing new methods for biomass conversion. Advanced biofuels, derived from biomass through innovative processes, offer a pathway to decarbonize the transportation sector while simultaneously providing a source of electricity generation. Integrating biomass energy into existing energy systems, such as using biogas to supplement natural gas supplies, offers a realistic path to enhance energy security and reduce reliance on fossil fuels. The development of sustainable biomass feedstock production systems, including the cultivation of dedicated energy crops, is crucial to ensure that biomass energy remains a viable and environmentally sound option for long-term electricity generation. The ongoing development and implementation of sustainable practices are key to ensuring that biomass contributes significantly to a cleaner, more sustainable energy future.

6.5 Ocean Energy

Ocean energy, a vast and largely untapped resource, offers a compelling alternative to traditional energy sources. Harnessing the power of the oceans presents unique engineering challenges, but the potential rewards—clean, sustainable energy on a massive scale—are undeniable. Several methods exist, each with its own set of advantages and limitations. Wave energy converters, for example, utilize the rhythmic motion of waves to generate electricity. These devices, ranging from simple oscillating water columns to complex articulated systems, are constantly being refined to improve efficiency and survivability in harsh marine environments. Research into materials science and hydrodynamic modeling is crucial for optimizing wave energy capture and minimizing environmental impact. The design considerations are complex, encompassing aspects of structural integrity, corrosion resistance, and energy transfer mechanisms.
Tidal energy, another promising avenue, leverages the predictable rise and fall of ocean tides. Tidal barrages, which are dam-like structures built across estuaries or inlets, are a well-established technology. However, their construction can be disruptive to ecosystems, necessitating careful environmental impact assessments and mitigation strategies. Tidal stream generators, on the other hand, resemble underwater wind turbines and are positioned within tidal currents to extract kinetic energy. These are less impactful than barrages, but their deployment necessitates robust anchoring systems and materials able to withstand the constant pressure and corrosion of saltwater. Careful site selection, to maximize tidal energy potential and mini-

mize environmental disruption, is paramount.
Ocean thermal energy conversion (OTEC) exploits the temperature difference between the warm surface water and the cold deep ocean. This technology utilizes a closed-cycle system employing a working fluid with a low boiling point, which is vaporized by the warm surface water, driving a turbine to generate electricity. The cooled vapor then condenses using the cold deep ocean water, completing the cycle. The scale of OTEC plants is substantial, requiring significant upfront investment and sophisticated engineering to handle the considerable volumes of water involved. Open-cycle OTEC systems are also being developed, and they directly use the warm surface water to create steam, which can then be used to drive a turbine. This method offers the added benefit of producing fresh water as a byproduct.
While the development and deployment of ocean energy technologies are progressing rapidly, several obstacles remain. The unpredictable and often harsh marine environment poses significant engineering challenges, necessitating the development of durable and corrosion-resistant materials. Grid integration presents another hurdle; many ocean energy installations are situated far from the onshore electrical grid. Transmission lines, often laid across the seabed, are expensive to install and maintain. Furthermore, the high capital costs associated with many ocean energy technologies can be a barrier to entry for smaller developers. Addressing these challenges requires collaboration between engineers, scientists, policymakers, and investors to unlock the enormous potential of ocean energy. The future of this field is bright, with ongoing research into improved designs, innovative materials, and effective grid integration strategies paving the way towards widespread adoption. The environmental benefits, combined with the potential for increased energy independence, make ocean energy a vital component of a sustainable energy future. Further research and development, coupled with supportive government policies, will be critical to fully realizing its potential and integrating it seamlessly into the global energy mix. The economic aspects, including the creation of new jobs and industries, are equally important and will be instrumental in driving further progress. This

technology's long-term viability depends on optimizing the economic model and ensuring financial sustainability for projects. Continued advancement in technology, including the development of more efficient and cost-effective devices, will be essential for wider market penetration and acceptance. Addressing the environmental concerns through responsible site selection and mitigation strategies is critical for building public trust and acceptance. Ultimately, the successful integration of ocean energy into the global energy system requires a multifaceted approach encompassing engineering excellence, strategic investment, and environmental stewardship.

7 Chapter 7: The Smart Grid

Electricity's journey from generation to consumption has dramatically changed. No longer is it a simple linear process; instead, we're navigating the complexities of a dynamic, interconnected network. This network, the smart grid, is a critical component of the future of power, a far cry from the rudimentary systems of our electrical pioneers.

The traditional power grid, a largely static system, is being fundamentally reshaped. This transformation involves integrating advanced technologies to improve efficiency, reliability, and resilience. We're talking about real-time monitoring, data analytics, and sophisticated control systems, all working in concert to manage the flow of electricity far more effectively than ever before. Think of it as moving from a basic, one-way street to a complex highway system with multiple lanes, sensors, and intelligent traffic control.

This modernization isn't just about upgrading existing infrastructure; it's about a complete overhaul of how we think about and manage electricity. Imagine a grid that can automatically reroute power around outages, preventing widespread blackouts. This is the potential of the smart grid – to anticipate and adapt to changing demands. Such adaptability is crucial as we incorporate more variable renewable sources like solar and wind power into our energy mix. These sources, while clean and sustainable, are inherently intermittent, presenting unique challenges to the stability of the grid.

Central to this shift is the implementation of advanced digital technologies. Smart meters, for example, provide real-time data on energy consumption, allowing utilities to better understand demand patterns and optimize resource allocation. This data-driven approach allows for more precise forecasting, leading to reduced waste and improved cost-effectiveness. This is not just a convenience, it's a fundamental shift in the way we manage this crucial resource. Think about the implications for energy efficiency campaigns, the precise targeting of incentives, and even the optimization of energy pricing strategies based on real-time demand.

Furthermore, the incorporation of energy storage solutions is paramount to the success of the smart grid. Batteries, pumped hydro storage, and other storage technologies play a vital role in smoothing out the fluctuations inherent in renewable energy generation. These systems act as buffers, storing excess energy during periods of high production and releasing it when demand is high or renewable generation is low. This capability is instrumental in ensuring grid stability and reliability, even with the increased penetration of intermittent renewable sources. The benefits extend beyond simply maintaining the lights on; these solutions play a crucial role in managing peak demand, optimizing energy prices, and improving the overall resilience of the electricity grid.

The challenge, however, lies in the significant investment required for grid modernization. Updating infrastructure, implementing advanced technologies, and training personnel are all costly endeavors. Yet, the long-term benefits far outweigh the initial costs. A more efficient, reliable, and resilient grid translates to lower energy prices, reduced greenhouse gas emissions, and a more sustainable energy future. This represents a significant undertaking, requiring collaboration between government, industry, and research institutions. Strategic planning, effective policy frameworks, and investments in research and development are all essential to unlock the full potential of the smart grid.

However, the transition to a smart grid also presents opportunities for innovation and economic growth. The development and deployment of new technologies create jobs, stimulate economic activity, and promote technological advancements across numerous sectors. The development of smart grid technologies represents a significant area of innovation, driving advancements in areas such as data analytics, sensor technology, communication networks, and energy storage. This is a growth area with vast potential for entrepreneurs, investors, and researchers alike. This isn't merely about upgrading the electricity grid; it's about building a smarter, more efficient, and sustainable future.

The smart grid isn't just a technological upgrade; it's a paradigm shift in how we manage and consume energy. This transformation requires a multifaceted approach, encompassing technological advancements, policy initiatives, and public awareness. The successful implementation of the smart grid will be essential in enabling the broader adoption of renewable energy sources and achieving a sustainable energy future. We are not merely improving an existing system; we are creating a new energy ecosystem.

7.1 Grid Modernization

Grid modernization is crucial for the efficient and reliable delivery of electricity in the 21st century. Our existing grids, often built decades ago, struggle to integrate the intermittent nature of renewable energy sources like solar and wind power. This necessitates a significant upgrade, moving beyond traditional infrastructure to a more dynamic and responsive system.

Modernizing the grid involves several key components. Advanced sensors and communication networks provide real-time data on energy flow, enabling predictive maintenance and preventing outages before they occur. This proactive approach minimizes disruption and maximizes the lifespan of critical grid infrastructure. Imagine the impact: fewer power failures, less wasted energy, and a more resilient system capable of handling unexpected surges in demand.

Smart meters, deployed at the consumer level, allow for two-way communication between the utility and the end-user. This bidirectional data flow enables time-of-use pricing, incentivizing consumers to shift their energy consumption to off-peak hours. Such dynamic pricing models help to balance energy supply and demand, smoothing out fluctuations caused by intermittent renewables. Furthermore, it empowers consumers to actively participate in managing the grid, fostering a more collaborative and sustainable energy ecosystem.

The incorporation of advanced analytics into grid management significantly improves operational efficiency. Sophisticated algorithms process vast amounts of data from various sources – weather patterns, energy consumption trends, and generator performance – to optimize energy distribution and predict potential problems. This data-driven approach allows for more efficient resource allocation, reducing energy waste and improving overall grid performance. The implications are far-reaching, leading to lower costs and increased reliability for both utilities and consumers.

Energy storage technologies are rapidly evolving and becoming increasingly integrated into modernized grids. Batteries, pumped hydro storage, and compressed air energy storage can help to mitigate the intermittency of renewable energy sources. By storing excess energy generated during periods of high production (like sunny midday hours), these systems ensure a consistent power supply even when renewable sources are unavailable. This smooths out the energy supply, improving grid stability and resilience and minimizing reliance on fossil fuel-based backup generators.

Cybersecurity is paramount in the context of a modernized, interconnected grid. Protecting the grid from cyberattacks is crucial, given its critical role in modern society. Investing in robust cybersecurity measures is not merely a cost but a necessity for maintaining grid reliability and preventing widespread disruption. The potential consequences of a successful cyberattack on the energy grid are simply too significant to ignore; proactive measures are paramount.

The successful modernization of the grid also depends on collaboration across various stakeholders. Utilities, government agencies, technology providers, and consumers must work together to create a sustainable, reliable, and efficient energy future. This collaborative effort will not only transform the energy landscape but also create numerous economic opportunities in research, development, and deployment of new technologies. The collaborative approach ensures that modernization initiatives are aligned with broader societal goals. Investing in the workforce is critical for the success of grid modernization projects. Training programs are essential to equip technicians and engineers with the necessary skills to operate and maintain the advanced technologies incorporated into the smart grid. This targeted training ensures that individuals can handle challenges like data analytics, cybersecurity threats, and the integration of renewable energy sources. A skilled and capable workforce is essential for driving the modernization process forward successfully.

The transition to a modernized grid is not without its challenges. Upgrading aging infrastructure requires significant capital investment, and integrating new technologies can be complex. However, the long-term benefits of increased efficiency, reliability, and sustainability far outweigh these initial costs. Moreover, the transition itself offers opportunities for innovation, economic growth, and the creation of new jobs in emerging sectors.

The benefits of a modernized grid extend beyond simple efficiency gains. It creates a platform for further advancements in energy technologies, fostering innovation and competitiveness in the energy sector. It allows for greater integration of distributed generation, empowering consumers and businesses to become active participants in the energy system. It is a critical step toward a sustainable energy future, enabling greater penetration of renewable energy sources and minimizing the environmental impact of energy production and distribution. Ultimately, a modernized grid represents a significant leap towards a cleaner, more reliable, and more resilient energy system for generations to come.

7.2 Digital Grid Management

Digital grid management represents a paradigm shift in how we control and optimize our power systems. Gone are the days of solely reactive grid operations; now, we leverage advanced technologies to anticipate and respond to changes in real-time, improving efficiency and reliability across the entire network. This sophisticated approach relies heavily on data analytics and predictive modeling, allowing operators to proactively address potential issues before they escalate into widespread outages.

The heart of digital grid management lies in its ability to monitor and control every aspect of the power system. Sensors strategically placed throughout the grid collect vast amounts of data—voltage levels, current flows, frequency fluctuations, and more. This data feeds into sophisticated algorithms that analyze the network's performance and predict future behavior. Identifying potential bottlenecks, forecasting energy demand, and optimizing power distribution become significantly more accurate and efficient with this constant stream of information. Imagine a network of interconnected brains constantly monitoring and adjusting the flow of electricity, ensuring a seamless supply across vast distances.

This predictive capability leads to enhanced grid stability. By anticipating fluctuations in energy demand or identifying potential fault locations, operators can proactively adjust power flows, preventing cascading failures and minimizing disruptions to consumers. This proactive approach contrasts sharply with the largely reactive strategies of the past, where outages were often addressed only after they had already occurred, leading to significant inconvenience and economic losses. The predictive modeling, fueled by real-time data analysis, minimizes downtime and allows for smoother, more reliable power delivery.

Furthermore, digital grid management facilitates the integration of renewable energy sources. The intermittent nature of solar and wind power presents unique challenges for grid stability, and integrating these sources requires careful coordination. Smart grid technologies allow for the seamless integration of these renewable resources by dynamically adjusting power flows in response to their fluctuating output. This is particularly crucial as we transition towards a future reliant on sustainable energy sources. The algorithms consider the intermittent nature of renewables, forecasting their output and adjusting the grid accordingly to prevent instability.

Advanced data analytics are also key to identifying and resolving inefficiencies within the power grid. By analyzing consumption patterns, operators can identify areas where energy is wasted or where infrastructure upgrades are needed. This targeted approach to grid optimization reduces operational costs and improves overall system efficiency, ultimately benefiting both consumers and energy providers. Identifying areas of high energy loss through detailed analysis leads to more efficient and targeted upgrades, optimizing the distribution process. These aren't just theoretical improvements; they translate into tangible economic savings.

However, the implementation of digital grid management comes with its own set of challenges. The sheer volume of data generated by the grid requires robust data management systems and sophisticated analytical tools. Security is paramount; ensuring the integrity and confidentiality of this data is critical to prevent cyberattacks and system failures. The cost of upgrading existing infrastructure to accommodate smart grid technologies can be substantial. It's not simply a matter of installing new equipment; it involves integrating complex systems, training personnel, and managing significant data streams.

The future of digital grid management involves ongoing development and refinement. Artificial intelligence and machine learning are playing an increasingly important role in enhancing grid intelligence, allowing for even more accurate predictions and automated responses to system events. The integration of advanced communication technologies, like 5G, promises further improvements in data transmission speeds and reliability. Advanced grid control systems can enhance efficiency, reducing energy losses and improving overall grid performance. Continuous development is paramount to maintain the grid's security and resilience in the face of ever-evolving challenges. The digital grid is not a static system; it's a dynamic entity constantly adapting to meet our evolving energy demands. Investing in these improvements guarantees a more stable and reliable power supply.

7.3 Energy Storage Solutions

Energy storage is no longer a futuristic concept; it's a crucial component of a reliable and efficient modern power grid. As renewable energy sources like solar and wind become increasingly prevalent, the intermittent nature of their power output presents a significant challenge. Unlike traditional power plants that can adjust their output on demand, solar panels only produce electricity when the sun shines, and wind turbines only generate power when the wind blows. This variability necessitates robust energy storage solutions to bridge the gap between energy supply and demand, ensuring consistent power delivery.

The most widely discussed and utilized energy storage solution is pumped hydro storage. This technology uses excess electricity to pump water uphill into a reservoir, then releases that water through turbines during periods of high demand, generating electricity. It's a mature, reliable technology, but its geographical limitations—requiring specific topography—restrict its widespread applicability. Furthermore, the environmental impacts, including habitat disruption and water usage, should be carefully considered in planning and implementation.

Beyond pumped hydro, several other promising energy storage technologies are rapidly developing. Batteries, particularly lithium-ion batteries, are seeing significant advancements, offering higher energy densities and longer lifespans. Their use in electric vehicles is driving down costs and improving performance, making them increasingly attractive for grid-scale applications. However, challenges remain, including the sourcing of critical raw materials and the environmental concerns associated with battery manufacturing and disposal. Research into alternative battery chemistries, such as solid-state batteries, aims to address these limitations.

Thermal energy storage systems offer another avenue for grid stabilization. These systems store energy as heat, often using molten salts or other materials with high thermal capacity. They are particularly well-suited for integrating concentrated solar power plants, where solar energy is used to heat a fluid, which is then stored and used to generate electricity later. The efficiency and long-term storage capabilities of these systems are significant advantages, although the high initial investment costs and potential safety concerns need careful consideration.

Compressed air energy storage (CAES) represents a different approach. Excess electricity is used to compress air, which is then stored in underground caverns or depleted gas fields. When energy is needed, the compressed air is released to drive turbines and generate electricity. CAES offers significant storage capacity, but its geographical limitations and the energy losses during compression and decompression processes limit its overall efficiency. However, improvements in technology are continually addressing these issues, potentially expanding its future role in the grid.

Flywheel energy storage employs rotating mechanical devices to store kinetic energy. These systems offer fast response times and high efficiency, making them suitable for applications requiring rapid power delivery, such as stabilizing grid frequency fluctuations. However, their energy density is relatively low compared to other technologies, limiting their overall storage capacity. Ongoing research focuses on developing higher-strength materials and improved designs to increase their energy density and expand their applications.

Ultimately, the optimal energy storage solution will depend on several factors, including the specific energy needs, geographical location, available resources, and economic considerations. A diverse portfolio of energy storage technologies is likely to be necessary to meet the growing demands of a decarbonizing energy landscape. The continued development and deployment of these technologies will be crucial for ensuring a reliable, sustainable, and efficient energy future, minimizing disruptions caused by the intermittency of renewable sources and ensuring a consistent power supply. This diversity and adaptability are vital for navigating the complexities of a constantly evolving energy sector. Further investment in research and development of each of these methodologies will further expand the effectiveness and reliability of the solutions available. We must also consider the implications for the workforce and how training opportunities must keep pace with advancements in the sector. A concerted effort towards this ensures that we harness these solutions optimally, while adapting to the ever-changing nature of energy demands. The successful integration of energy storage solutions is not merely a technological challenge, but also a complex economic and societal undertaking.